Thin-Film Design

Modulated Thickness and
Other Stopband Design Methods

Tutorial Texts Series

Thin-Film Design

Modulated Thickness and Other Stopband Design Methods

Bruce E. Perilloux

Tutorial Texts in Optical Engineering
Volume TT57

Arthur R. Weeks, Jr., Series Editor
Invivo Research Inc. and University of Central Florida

SPIE PRESS
A Publication of SPIE—The International Society for Optical Engineering
Bellingham, Washington USA

Library of Congress Cataloging-in-Publication Data

Perilloux, Bruce.
    Thin-film design : modulated thickness and other stopband design methods / by Bruce
    Perilloux.
        p. cm.—(SPIE Press tutorial texts ; v. TT 57)
    Includes bibliographical references and index.
    ISBN 0-8194-4525-8 (softcover)
    1. Thin films. I. Title. II. Tutorial texts in optical engineering ; v. TT 57.

TA418.9.T45 P47 2002
621.3815'2—dc21   2002019550
                                                                 CIP

Published by

SPIE—The International Society for Optical Engineering
P.O. Box 10
Bellingham, Washington 98227-0010 USA
phone: 360.676.3290
fax: 360.647.1445
email: spie@spie.org
www.spie.org

Introduction to the Series

Since its conception in 1989, the Tutorial Texts series has grown to more than 60 titles covering many diverse fields of science and engineering. When the series was started, the goal of the series was to provide a way to make the material presented in SPIE short courses available to those who could not attend, and to provide a reference text for those who could. Many of the texts in this series are generated from notes that were presented during these short courses. But as stand-alone documents, short course notes do not generally serve the student or reader well. Short course notes typically are developed on the assumption that supporting material will be presented verbally to complement the notes, which are generally written in summary form to highlight key technical topics and therefore are not intended as stand-alone documents. Additionally, the figures, tables, and other graphically formatted information accompanying the notes require the further explanation given during the instructor's lecture. Thus, by adding the appropriate detail presented during the lecture, the course material can be read and used independently in a tutorial fashion.

What separates the books in this series from other technical monographs and textbooks is the way in which the material is presented. To keep in line with the tutorial nature of the series, many of the topics presented in these texts are followed by detailed examples that further explain the concepts presented. Many pictures and illustrations are included with each text and, where appropriate, tabular reference data are also included.

The topics within the series have grown from the initial areas of geometrical optics, optical detectors, and image processing to include the emerging fields of nanotechnology, biomedical optics, and micromachining. When a proposal for a text is received, each proposal is evaluated to determine the relevance of the proposed topic. This initial reviewing process has been very helpful to authors in identifying, early in the writing process, the need for additional material or other changes in approach that would serve to strengthen the text. Once a manuscript is completed, it is peer reviewed to ensure that chapters communicate accurately the essential ingredients of the processes and technologies under discussion.

It is my goal to maintain the style and quality of books in the series, and to further expand the topic areas to include new emerging fields as they become of interest to our reading audience.

Arthur R. Weeks, Jr.
Invivo Research Inc. and University of Central Florida

Contents

Chapter 3   Discrete Apodization of TMDs / 45

Chapter 4   Other Complex TMDs / 61

Chapter 5   Quarter-wave Stack Transformation / 75

Appendix A: Useful Equations for Discrete-layer Thin-Film Calculations / 91

Appendix B: Chebyshev Polynomials of the Second Kind / 93

Appendix C: FORTRAN90 Source Code for the Determination of All Possible TMD Stopbands / 95

Preface

This text is written for thin-film designers and students with advanced knowledge of multilayer, optical thin-film coatings. It focuses specifically on coatings that have high reflectance performance requirements in more than one spectral wavelength band or region. Many advanced optical systems that employ optical thin-film coatings rely on the performance attributes of multiple spectral bands. Several new analytical design methods that produce multiple stopbands as well as passbands are presented in this text. These analytical design methods produce discrete thin-film designs that will, in most cases, achieve specifications without any refinement of the design. If needed, the spectral performance of these analytical designs could be improved via common computer refinement algorithms. The theory of each design method in this text is presented along with design examples. Relatively basic exercises are provided for students as well as challenging ones for researchers.

This text does not attempt to cover the vast amount of material already published on thin films. The reader is expected to have a general knowledge of classical thin-film physics and designs. Several other texts cover the gamut of classical thin-film theory and designs, and additional information is readily available in several conference proceedings and many journal articles. Detailed derivations of Fresnel equations, complex refractive index, effective index, p- and s-polarization, phase, classical thin-film coating designs, etc., were intentionally omitted from this text. However, a brief introductory chapter on thin-film theory and design is included for completeness.

The coating designs that readers produce using the methods described in this text can be readily manufactured using common coating materials and process methods. In general, the layer thicknesses of these designs vary from a quarter-wave stack to a modulated profile. For some designs, process adjustments may be required for thin layers. However, thin-film manufacturing processes are not covered in this text since other texts and technical papers cover them.

The designs produced using the methods in this text can also be used as initial designs for computer refinement and synthesis algorithms. With some commercially available thin-film design software, some designs are produced with no starting design or specified layers; the algorithm adds layers of preselected materials until the desired spectral performance is achieved. Still, a good starting design helps to reduce the time or effort required to produce a final design that (1) achieves desired performance specifications; (2) is insensitive to layer thickness errors; and (3) can be manufactured. The design methods presented here are also expected to accomplish these tasks for some applications that require stopbands.

In general, the analytical design methods presented were developed using the following methodology. First, layer thicknesses of an arbitrary quarter-wave stack were modulated using various mathematical functions (e.g., sinusoidal). Next, a computer program was written to determine the existence of all stopbands produced from the modulated design. The resulting patterns of stopbands were evaluated graphically as functions of modulation parameters and spectral frequency.

Then, based on these graphical patterns of stopbands, analytical (linear) equations were tested by direct calculation of spectral performance to see if the stopbands could be reproduced analytically.

Several variations on the modulation of layer thickness are presented in this text, including an inhomogeneous rugate design. The last chapter presents a related, novel design method where one quarter-wave stack is linearly transformed into another. Here, empirical testing of layer thickness was used to develop general transform equations. A summary of each chapter and the appendixes follows:

Chapter 1 reviews the fundamental mathematics for thin-film design that applies to the proposed methods. Again, the objective here was to keep this chapter brief since this information can be found in many texts.

Chapter 2 introduces sinusoidal thickness modulation of quarter-wave stacks. First, stopbands and passbands are defined. Next, modulation parameters are assessed and many designs are evaluated. Then linear equations are determined that predict all possible stopbands. The last section evaluates the electric fields and the reflected differential phase shifts of some modulated designs.

Chapter 3 introduces discrete apodization of the modulated designs from Chapter 2. Two specific apodization functions are evaluated: amplitude modulation functions and Gaussian envelope functions. Linear equations are again determined that predict all possible stopbands.

Chapter 4 describes two variations of the modulation scheme from Chapter 3. First, chirped-modulation designs are evaluated for spectral performance. Next, a half-modulation is discussed where every other layer of a quarter-wave stack is modulated. Both of these methods are applied to dispersion-controlled mirrors used to produce femtosecond laser pulses. Two design examples and the limitations of these modulation schemes are covered.

Chapter 5 presents a novel, linear transform method that can be used to partially transform a given quarter-wave stack into a second quarter-wave stack. This transformation is accomplished by adjusting the individual layer thickness while the total thickness of the original quarter-wave stack remains constant. General transform equations are developed by direct numerical testing of the transform method. The purpose of this transform is to obtain, or achieve, some of the spectral properties of both quarter-wave stacks (i.e., stopbands).

The five appendixes provide some useful thin-film equations, the Chebychev polynomials used in Chapter 2, the FORTRAN source code used to determine all possible stopbands of modulated designs, several graphs of stopband positions, and a summary of the linear equations that predict stopband positions and the general linear transform equations.

Hopefully this text will provide readers with some new thin-film design tools, further insight to design methods, and inspiration for further research on thin-film design.

I would like to thank my family for their support of my research and writing of this text. I greatly appreciate several helpful discussions with Dr. Philip Baumeister, Dr. Angus Macleod, and Dennis Fischer on these modulation design methods. I also acknowledge the collaborative investigation of rugate versions of these modulated designs, and the rugate designs and calculations, provided by Dr. Pierre Verly at the National Research Council Canada. I would also like to thank the SPIE reviewers for suggesting several improvements to this text. I would like to thank Coherent, Inc. for supporting this work. Lastly, I would like to acknowledge my graduate professor, Dr. Rasheed M. A. Azzam, for his inspiration to continue my research of thin films.

Bruce E. Perilloux
September 2002

Definitions

Additive fractional units (S^+) The number of fractional parts of a quarter-wave layer(s) that combine to produce a new quarter-wave layer, as part of a new quarter-wave stack (see *fractional units*).

Amplitude-modulated TMD (AM-TMD)   A variation of a thickness-modulated design where the modulation amplitude of the TMD is also modulated at a lower frequency.

Apodization   Envelope function of refractive-index profiles of rugate filters used to suppress passband ripples; also used as envelope functions for layer thickness of TMDs.

Base period (TMD)   The integer number of layers that have a nonrepeating thickness pattern (see *modulation period*).

Chirped TMD (C-TMD)   A thickness modulated design where the modulation frequency varies or changes as a function of layer number.

Degenerate TMD   Typically, a thickness modulated design that has no modulation (i.e., a quarter-wave stack).

Fractional units (U) The integer number of fractional parts that a quarter-wave thick layer must be subdivided into in order to transform a quarter-wave stack into a second stack (see *LOST equations*).

Gaussian envelope function TMD (G-TMD)   A thickness-modulated design with the modulation amplitude determined by a Gaussian function.

Half-modulated TMD (H-TMD)   A variation of a thickness-modulated design where the modulation is applied to every other layer; the layers that are not modulated all have the same optical thickness.

HWOT   Half-wave optical thickness.

Linear optical stack transform (LOST)   A linear transform of a quarter-wave stack where the layer thicknesses are adjusted to produce a second quarter-wave stack; a partial transform results in some properties from both quarter-wave stacks.

LOST equations   Two general equations used to linearly transform the additive and subtractive fractional units of the quarter-wave layers of one quarter-wave stack into a second quarter-wave stack.

Modulation amplitude (k) The amplitude at which layer thicknesses are modulated (0–1).

Modulation frequency (f) The frequency at which layer thicknesses are modulated.

Modulation period (T) The number of layers of a modulated design that produces a unique layer pattern.

Quarter-wave stack transform   See *linear optical stack transform*.

QWOT   Quarter-wave optical thickness.

Rugate TMD   An inhomogeneous design in which the continuously refractive index profile is determined using a Fourier transform of the desired transmission produced from a discrete-layered thickness-modulated design.

Subtractive fractional units (S^-) The number of fractional parts of a quarter-wave layer(s) that is subtracted from additive fractional units to produce a new quarter-wave layer, as part of a new quarter-wave stack (see *fractional units*).

Thickness-modulated design (TMD)   A quarter-wave stack that has the optical thickness (QWOT) of its layers modulated by a periodic function, for example, a sinusoidal function.

Universal stopband equation (USE)   A linear equation that is a function of modulation frequency, which predicts all possible stopbands for thickness-modulated designs.

Chapter 1
Introduction

The process of designing a thin-film coating can be approached from an analytical or a scientific basis, a numerical or an actual experience, or—with the capability of present thin-film design software—perhaps with no basis (knowledge) at all. Over many decades, several design, analysis, and other related books and technical articles have been published on this subject. As more optical systems have become multispectral with performance requirements at several wavelengths or wavelength regions, the required coatings have become more complex. A particular subset or type of coating design is typically used to highly reflect at several wavelengths and may also have high transmittance at others. Classical designs, such as bandpass, cavity, notch, minus filters, rugate, etc., are routinely used for these applications. This chapter is a brief review of well-established thin-film theory as it relates specifically to the above subset of coating designs.

1.1 Review of Mathematics for Thin-Film Design

The basic physics of optical thin films involves the interaction of one or more wavelengths of light and real media boundaries. Without these boundaries, Maxwell's equations describe properties of light waves that are typically modeled as planar or Gaussian. With boundary conditions, where different dielectric or conductive media are present, Maxwell's equations are solved accordingly to describe the reflected and transmitted waves. Specifically, the complex reflection coefficient is the amplitude ratio of the reflected to incident electric fields. From this coefficient, the reflectance (which is the ratio of the reflected wave's power to the incident wave's power) and the phase shift between the reflected and incident waves are determined. The rigorous mathematical development of the electromagnetic theory for propagating waves and interactions with boundaries for various materials is covered in other texts on electromagnetics[1] and optics.[2]

Beyond the fundamental electromagnetic concepts of propagating waves or beams and media and boundary conditions, single or multiple layers of so-called optical "thin" films have two or more boundaries with relatively small spacing or distance between the boundaries (i.e., this spacing is approximately on the order of the wavelength of the light source). These boundaries or interfaces cause a plurality of reflected waves that can interfere with each other in amplitude and phase. These reflected waves interfere, or add coherently, because the path differences within thin films are less than the coherence length, given by Eq. (1.1):[3]

$$l \approx \frac{\lambda^2}{\Delta\lambda}. \tag{1.1}$$

Here, l is the coherence length, λ is the wavelength of light, and $\Delta\lambda$ is the bandwidth. Over the short distances of thin films, most light sources are coherent and interference is observed. For larger distances (e.g., a millimeter) and for quasi-monochromatic light sources, interference can also be observed (the path length is still less than the coherence length). However, for path differences larger than the coherence length, multiple waves interfere with each other in intensity only (in an incoherent manner).[3] Still, the observation of interference effects depends on the detection conditions. This text investigates thin-film designs based on the optical interference of thin films, where an incident wave of light is multiplied into groups of reflected and transmitted waves.

The spectral performance of optical thin-film designs can be computer-modeled using any one of four methods or models (electric field matrix, characteristic matrix, reflection recursion, and admittance recursion).[4] For the purpose of this text, the characteristic matrix is used to model thin-film designs for the particular case of nonabsorbing media. The derivation of the characteristic matrix is not included because it is thoroughly covered in other thin-film texts. However, for completeness, definitions of characteristic matrix, admittance, and the calculation of reflectance are provided.[5]

From Ref. [5], the characteristic matrix for any isotropic and homogeneous thin-film layer is given by

$$\begin{bmatrix} \cos\delta_L & (i\sin\delta_L)/\eta_L \\ i\eta_L\sin\delta_L & \cos\delta_L \end{bmatrix}, \tag{1.2}$$

where the phase term δ_L is given by

$$\delta_L = \frac{2\pi N_L t_L}{\lambda}\cos(\theta_L). \tag{1.3}$$

N_L is the complex refractive index of the layer given by

$$N_L = n_L - ik_L, \tag{1.4}$$

where n_L and k_L are the refractive index and extinction coefficient, respectively, and t_L is the layer's physical thickness. Using Snell's law, the internal angle θ_L is readily found from the incident angle of the beam θ_0, from the complex refractive index of the incident medium N_0, and from N_L, where

$$N_0\sin(\theta_0) = N_L\sin(\theta_L). \tag{1.5}$$

The optical admittance of the layer, η_L, is dependent on the polarization of the incident beam and is given by

$$\text{s-polarization:} \quad \eta_L = N_L\cos(\theta_L) \tag{1.6}$$

and

$$\text{p-polarization:} \quad \eta_L = N_L/\cos(\theta_L). \tag{1.7}$$

It should be noted that the term for optical admittance of free space was omitted from Eqs. (1.6) and (1.7) because it cancels when solving for the admittance of the film-substrate optical system at the interface of the ambient medium and film. For reference, the optical admittance of free space Y_0 is the reciprocal of its optical impedance of free space Z_0 given by

$$Y_0 = \frac{1}{Z_0} = \sqrt{\frac{\varepsilon_0}{\mu_0}}, \qquad (1.8)$$

where ε_0 and μ_0 are the permittivity and permeability of free space, respectively.

The general form of the optical system described has an ambient medium and thin-film layer(s) deposited onto a substrate as shown in Fig. 1.1. A monochromatic light wave or beam in the ambient medium impinges on the thin-film layer(s) and substrate and is subsequently reflected and transmitted. By solving for the electric and magnetic field vector amplitudes at the ambient-film (outer layer #1, see Fig. 1.1) interface, the optical admittance for the film-substrate system is determined from its characteristic matrix. From this film-substrate's admittance and that of the ambient medium, the reflectance and transmittance are determined.

As one could infer from Fig. 1.1, a small lateral shift of a reflected beam could occur. Departing from plane wave theory, for the special case of total internal reflection (TIR), a shift does occur. This shift is called the Goos-Hänchen shift.[6] However, this shift is typically ignored in thin-film modeling.

Without going through the derivation, the general form of the characteristic matrix for a multilayer film-substrate system is given by the matrix

$$\begin{bmatrix} E_A \\ H_A \end{bmatrix}, \qquad (1.9)$$

where

$$\begin{bmatrix} E_A \\ H_A \end{bmatrix} = \left\{ \prod_{L=1}^{n} \begin{bmatrix} \cos \delta_L & (i \sin \delta_L)/\eta_L \\ i\eta_L \sin \delta_L & \cos \delta_L \end{bmatrix} \right\} \begin{bmatrix} 1 \\ \eta_S \end{bmatrix} \qquad (1.10)$$

and η_S is the admittance of the substrate medium given by

$$\eta_S = \frac{H_S}{E_S}. \qquad (1.11)$$

As shown in Eq. (1.10), the corresponding multilayer thin film-system matrix is the product of the characteristic 2×2 matrices for each layer in the system. The subscript L denotes the Lth layer, where $L = 1$ is the adjacent layer to the ambient medium. A multiplicative factor is not shown in Eq. (1.10) because it is not required to calculate the optical admittance of the film-substrate system as defined below.

The optical admittance of the film-substrate system is

$$Y = \frac{H_A}{E_A}. \tag{1.12}$$

The resulting product matrix of the 2×2 characteristic matrices from Eq. (1.10) takes the general form

$$\begin{bmatrix} p_{11} & p_{12} \\ p_{21} & p_{22} \end{bmatrix}. \tag{1.13}$$

By substituting Eq. (1.13) into Eq. (1.10) and using Eq. (1.12), the optical admittance Y for the film-substrate system can be determined, where

$$Y = \frac{p_{21} + p_{22}\eta_s}{p_{11} + p_{12}\eta_s}. \tag{1.14}$$

From the admittance Y of the ambient-coating-substrate system, the complex amplitude reflection coefficient r is determined from Eq. (1.14), where

$$r = \frac{\eta_A - Y}{\eta_A + Y}. \tag{1.15}$$

The optical admittance of the ambient medium, η_A, is determined from Eq. (1.6) or (1.7). Again, note that the optical admittance for each thin-film layer and ambient and substrate media are determined from Eqs. (1.6) and (1.7), and they depend on the incidence angle and polarization state of the incident beam.

The ratio of the reflected beam's power to that of the incident beam, or the reflectance, is given by

$$R = r \times r^*. \tag{1.16}$$

In the case of a nonabsorbing, scatter-free coating-substrate system, the transmittance is determined by

$$T = 1 - R. \tag{1.17}$$

The reflectance can be determined from Eq. (1.15) for the p- and s-polarization states, namely r_p and r_s, respectively. At oblique incidence angles, the differential phase shift Δ is determined from

$$\Delta = \delta_p - \delta_s, \tag{1.18}$$

where

$$\delta_{p,s} = \arg(r_{p,s}). \tag{1.19}$$

At normal incidence, the differential phase shift is either zero or π radians depending on the choice of phase convention: either Abelès (thin film) or Nebraska-Muller[7] (ellipsometry), respectively.

1.2 Analytical, Discrete-layer, Thin-Film Designs

Several types of classical thin-film designs achieve prespecified spectral performances based on analytical equations or known layer sequences. Even without the use of a computer and an algorithm to find or optimize an initial coating design, these thin-film coatings are readily designed.[8] Some very basic types of analytical coating designs are shown in Table 1.1.

The basic design types shown in Table 1.1 are written in common symbolic notation, where L, M, and H represent low, medium, and high refractive index layers, respectively. These layers are specified as quarter-wave optical thickness (QWOT), where

$$T_L = 4N_L t_L \cos \theta_L. \tag{1.20}$$

T_L is the thickness (QWOT) of a layer with refractive index N_L, physical thickness t_L, and incidence angle θ_L. Refractive-index profile plots show the refractive index of discrete film layers as a function of optical thickness, as shown in Figs. 1.2(a)–(e).

Table 1.2 lists some specific design types that are similar to designs presented in this text.

Figure 1.2 shows refractive-index profile plots for several of these designs. Also shown are optical-thickness (QWOT) profile plots as a function of the layer

Table 1.1 Some basic thin-film designs.

Coating	Common Purpose or Use	Some Basic Design Types
Antireflection (AR)	Windows	Ambient/L/Sub Ambient/LML/Sub Ambient/LMHL/Sub
Partial reflector (PR)	Laser output couplers Beamsplitters	Ambient/xL (HL)^n/Sub
High reflector (HR)	Laser cavity reflectors Corner mirrors	Ambient/(HL)^n/Sub
Polarizers	Separate polarization states	Ambient /(HL)^n/Sub
Dichroic filters	Separate wavelength bands Color filters	Ambient/(0.5L H 0.5L)^n/Sub Ambient/(0.5H L 0.5H)^n/Sub
Bandpass filters	Wavelength separation	Ambient/(LH)^n 2L (HL)^n/Sub

Table 1.2 Selected thin-film designs.

Coating	Some Basic Design Types	Typical Variations
High reflector (HR)	Ambient /(HL)^n/Sub	High-order quarter-wave stacks Multiple stacks Arithmetic or geometric stacks (Ref. [9])
Dichroic filters [short-wave pass (SWP), long-wave pass (LWP)]	Ambient /(L/2 H L/2)^n/Sub Ambient /(H/2 L H/2)^n/Sub	Combined SWP and LWP Nonpolarizing
Bandpass filters	Ambient /(LH)^n 2L(HL)^n/Sub	Single cavity Multiple cavity (Ref. [4])
Minus filters	Ambient /(aL/2 bH aL/2)^n/Sub where a + b = 2	Ref. [10]

number, which correspond to the refractive-index profiles. Refractive-index profile plots are useful to simultaneously visualize optical thickness and refractive index for individual discrete layers in a given coating design. In addition, electric-field amplitude curves are frequently overlaid on refractive-index profile plots, as shown in Fig. 1.2(a). However, for refined or complex analytical designs, optical-thickness profile plots can provide better visualization of layer thickness and layer thickness patterns than refractive-index profiles. The former is introduced and both are utilized in this text for the discrete layer designs presented.

The analytical designs shown in Table 1.2 and Fig. 1.2 are of interest in this text, but many other types of classical, analytical, and numerical coating designs are not covered. Instead, the concept of some specific analytical designs is included to show the classical designs that are related to the designs developed here. Next, we might ask the question "Is this design ready to be manufactured?" In some cases, analytical designs are not ready because of additional issues such as availability of film materials with the desired physical properties and dispersion of the selected materials. Typically the next step toward readying the design for manufacture is computer refinement.

1.3 Thin-Film Design Methods

This section first discusses the typical design process for thin-film coatings. A flow-chart for the design of a thin-film coating is shown in Fig. 1.3. It is of particular interest to look at how the preliminary design is determined. Here, the analytical path allows the designer to select from classical designs or to derive designs from *a priori* analytical methods. Alternatively, a design may already exist that meets or performs very close to the required specifications (empirical basis). Last, using computer software, the designer can refine or optimize an initial empirical or analytical starting design.[11] With newer algorithms and faster computers, the computer can then "create" a design; that is, given the layer, ambient and substrate materials, and spectral performance specifications, the algorithm can determine the number, thickness, and sequence of layers. However, it is debatable whether any, some, or all of the coating designs created by the latter process are "good" designs that are ready to be manufactured.

This text deals primarily with new analytical methods for discrete thin-film coatings along with some refinement of these initial designs. To provide insight and background information for some of the particular analytical methods covered, the last section in Chapter 1 covers rugate and synthesized rugate coatings. This discussion departs from the premise of this text on discrete thin-film designs, but rugate coatings and theory are closely related to the discrete-layer design methods presented.

1.4 Inhomogeneous Coating Designs and Synthesis

Early work on inhomogeneous thin films was done by Jacobsson et al.[12] The characteristic matrix for an inhomogeneous layer[13] is similar in form to Eq. (1.10) but rather complicated. Typically, inhomogeneous layers are computer-modeled by subdividing the given inhomogeneous layer into discrete, homogeneous ones.[14] Next, the quasi-exact spectral performance is readily found from Eq. (1.10). Furthermore, Ref. [14] describes how an inhomogeneous layer can be synthesized by optimizing the refractive index of the subdivided homogeneous layers to achieve a desired spectral performance (Snedaker). A brief review of rugate and synthesized rugate coatings is covered in Sec. 1.4.1. In addition to rugate designs, an inhomogeneous refractive-index profile can be determined for a given spectral curve or performance. The well-known Fourier transform method is presented in Sec. 1.4.2.

1.4.1 Rugate and synthesized rugate designs

Rugate coatings typically consist of a single layer deposited onto a substrate where the refractive index of the layer continuously changes as a function of depth. The nonabsorbing rugate film layer's refractive index is a function of the layer's thickness. A typical index profile of a rugate design is given in Eq. (1.21):[15]

$$n(x) = n_a + \frac{n_{pv}}{2} \sin\left(\frac{2\pi x}{n_a T} + \theta\right). \tag{1.21}$$

The basic sinusoidal refractive-index profile of a rugate filter (see Fig. 1.4) is determined from Eq. (1.21), where n_a and n_{pv} are the average and range of minimum-to-maximum refractive index, respectively; $n_a T$ is the period; θ is a phase term; and x is the optical distance from the substrate. The period of the sine wave is of interest since the wavelength of the single high-reflectance stopband produced with a rugate filter is determined. Specifically, the center wavelength of this stopband is twice the period, or

$$\lambda = 2n_a T, \tag{1.22}$$

where the period is defined in terms of optical thickness. By superimposing several sine waves for the refractive-index profile, a rugate coating produces several stopbands that correspond to each of the individual sine waves.

One particular enhancement of rugate coatings greatly reduces or suppresses reflection ripples or sidelobes in all passband wavelength regions, as shown in Fig. 1.5. This enhancement was defined by Southwell[16] as apodization, which utilizes envelope functions on sinusoidal thickness profiles to suppress sidebands. Apodization envelope functions can be Gaussian, sine, linear, or other functions.[15] An example of an envelope function from Ref. [11] is shown in Fig. 1.6 and applied to a sine wave-modulated index profile as determined by Eq. (1.21), where

$$n(x) = n_a + \left\{ \exp\left[-\beta \left(x - \frac{N n_a T}{2} \right)^2 \right] \frac{n_{pv}}{2} \sin\left(\frac{2\pi x}{n_a T} + \theta \right) \right\}. \qquad (1.23)$$

In Eq. (1.23), β is a phase term and N is the number of periods in the sine wave.

The use of wavelets was proposed by Southwell[17] to design several types of apodized rugate coatings. One application, discussed in more detail in Chapter 4, increases the bandwidth of reflector designs.

1.4.2 Fourier transform method of inhomogeneous coating design

Liddell[9] summarized the different types of exact and approximate Fourier transform synthesis methods. Dr. Pierre Verly developed a Fourier transform method and software for the design of the inhomogeneous coatings first proposed by Delano and Sossi.[18] As stated in Ref. [18], the analysis of a given inhomogeneous coating with the refractive-index profile $n(x)$ is exact: The analytical solution to the corresponding Maxwell's equations uniquely determines the spectral performance. A brief description of this Fourier transform method is given below.

To synthesize or determine the required inhomogeneous $n(x)$ profile that produces a known spectral performance, loss and dispersion are excluded. Based on these conditions, the $n(x)$ profile that produces a desired transmittance is given by[18]

$$n(x) = n_0 \exp\left[\frac{i}{\pi} \int_{-\infty}^{\infty} \frac{Q(T, k)}{k} e^{-ikx} dk \right], \qquad (1.24)$$

where $Q(T, k)$ is a complex function (commonly referred to as the Q-function), $k = 2\pi/\lambda$ is the wave number, λ is the wavelength, and x is twice the optical thickness. $T(k)$ is typically defined over a finite wavelength range and set to the value of 1 (unity) outside this range ($Q = 0$ when $T = 1$). Here, several forms of the Q-function have been used to overcome limitations, such as when the desired transmittance is near zero (high reflectance). It should be noted that it is necessary to neglect multiple reflections (multiple integrals) to obtain Eq. (1.24).[18] The National Research Council (NRC) of Canada, Bovard,[19] and others have reported several methods that solve for $n(x)$ for almost any given spectral performance, and even with high or low reflectance. For example, one Q-function useful for high reflectance was reported by Verly et al.,[20] where

$$|Q| = w\sqrt{1-T} + (1-w)\sqrt{\frac{1}{T}-1} \qquad (1.25)$$

and

$$0 \leq w \leq 1. \qquad (1.26)$$

In Eq. (1.25), T is the desired transmittance and w is an empirically determined weighting factor that is adjusted to achieve a best fit between the desired and calculated transmittance. It is well known that Q-functions are only approximate, and additional methods must be used to correct for the resulting errors. The phase of the Q-function is a function of both reflection and transmission phases, which are usually unknown. Depending on different transmission parameters, k, or the phase of the Q-function, the resulting refractive-index profile will be different.[21] Still, accurate solutions are readily obtained for most transmission parameters. Once the inhomogeneous $n(x)$ profile is determined, an approximate, discrete-layer design can be determined (i.e., discretization). A complete discussion of this method is provided in Refs. [18] and [20].

1.5 Summary

In this chapter, the mathematics for one method to model thin-film performance—namely, the characteristic matrix—was presented first, along with the optical admittance of a multilayer (general case) film-substrate system in order to determine reflectance and transmittance. This swift treatment of electromagnetic and thin-film physics was by no means an attempt to trivialize it; quite simply, the extensive details, development, special cases, etc., of thin-film theory are readily available in several texts. The above presentation's format quickly develops Eqs. (1.10) and (1.14)–(1.19) in a format that is readily usable for the computation of reflectance and transmittance of multilayer thin-film designs (see Appendix A for useful forms of these equations).

Next, discrete-layer thin-film coating design types were presented, with emphasis on dichroic, bandpass, notch, and band-reject filters. The general design method was included to complete the connection between analytical designs and computer refinement methods.

Lastly, in order to provide the reader with the essential background information necessary for the design methods presented, rugate and inhomogeneous coating design methods were reviewed. Although the design methods presented later in this text are specifically for discrete-layer designs, a close connection exists between these designs and inhomogeneous coating designs.

1.6 Exercises

1. Two polished, transparent glass monitor chips are stuck together or optically contacted. What is the maximum spacing between these chips (air gap) where interference patterns can be observed at $\lambda = 550$ nm? Assume $\Delta\lambda = 1$ nm.

2. An uncoated 1-mm thick ZnSe, plano/plano window ($n = 2.403$ at $\lambda = 10.6\,\mu$m) has a measured transmittance of 68.9% in air at normal incidence. Derive the equation for total transmittance (assuming optical interference between the two surfaces) and determine the total transmittance.

3. Starting with Eq. (1.10), where the electric and magnetic field vectors are determined at the surface boundary of a single layer stack, replace the optical admittance of the layer η with NY_0, where Y_0 and N are the optical admittance of free space and the complex refractive index of the layer, respectively. Solve for the optical admittance of the system, Y, for normal incidence.

4. Derive the optical admittance for a single, nonabsorbing layer for normal incidence and for arbitrary p- and s-polarizations.

5. Write a thin-film program that calculates reflectance and transmittance for any given incidence angle, polarization, ambient medium, multilayer thin-film design, and substrate.

6. Optimize any starting design (or no design) with two thin-film materials to produce a dual-wavelength high reflector, multilayer dielectric coating with stopbands at wavelengths 1064 nm and 780 nm, with reflectance greater than 99.8% at normal incidence. Assume nondispersive, nonabsorbing refractive indices of 1.0, 1.46, 2.25, and 1.52 for the ambient, the two layer materials, and the substrate. How many layers were required? What are the maximum and minimum layer thicknesses (if any) in the final design compared with the original design?

7. Synthesize a rugate design (homogeneous layers). Using thin-film software, set up 20 nondispersive coating material files with equally spaced refractive- index values between 1.46 and 2.25. Next, using Eqs. (1.21) and (1.22), determine the required layer thickness and period needed to produce a stopband at $\lambda = 1\,\mu$m with reflectance greater than 99%.

8. Plot the reflectance of the design from Exercise 5 within the wavelength range of $\lambda = 0.5 - 2\,\mu$m. Using Gaussian apodization [Eq. (1.23)] on the same design, plot the reflectance again and compare reflectance ripples in the passbands.

9. Explain why a Q-function would not produce unique solutions for a reflectance profile with high reflectance.

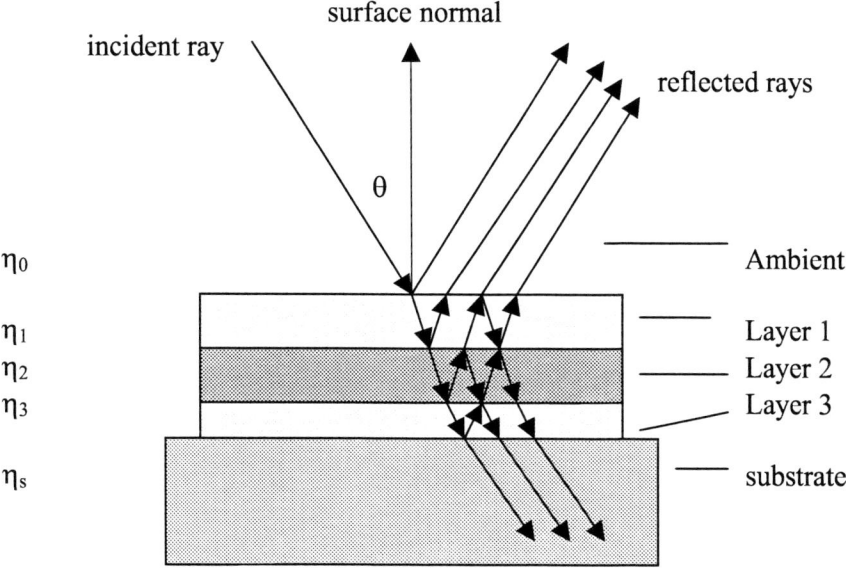

Figure 1.1 surface normal, incident ray, reflected rays, θ, η₀ Ambient, η₁ Layer 1, η₂ Layer 2, η₃ Layer 3, ηs substrate, refracted rays (transmitted)

Figure 1.1 Typical cross-section ray trace diagram for an arbitrary, multilayer thin film-substrate system.

Refractive-Index Profiles   Layer-Thickness Profiles

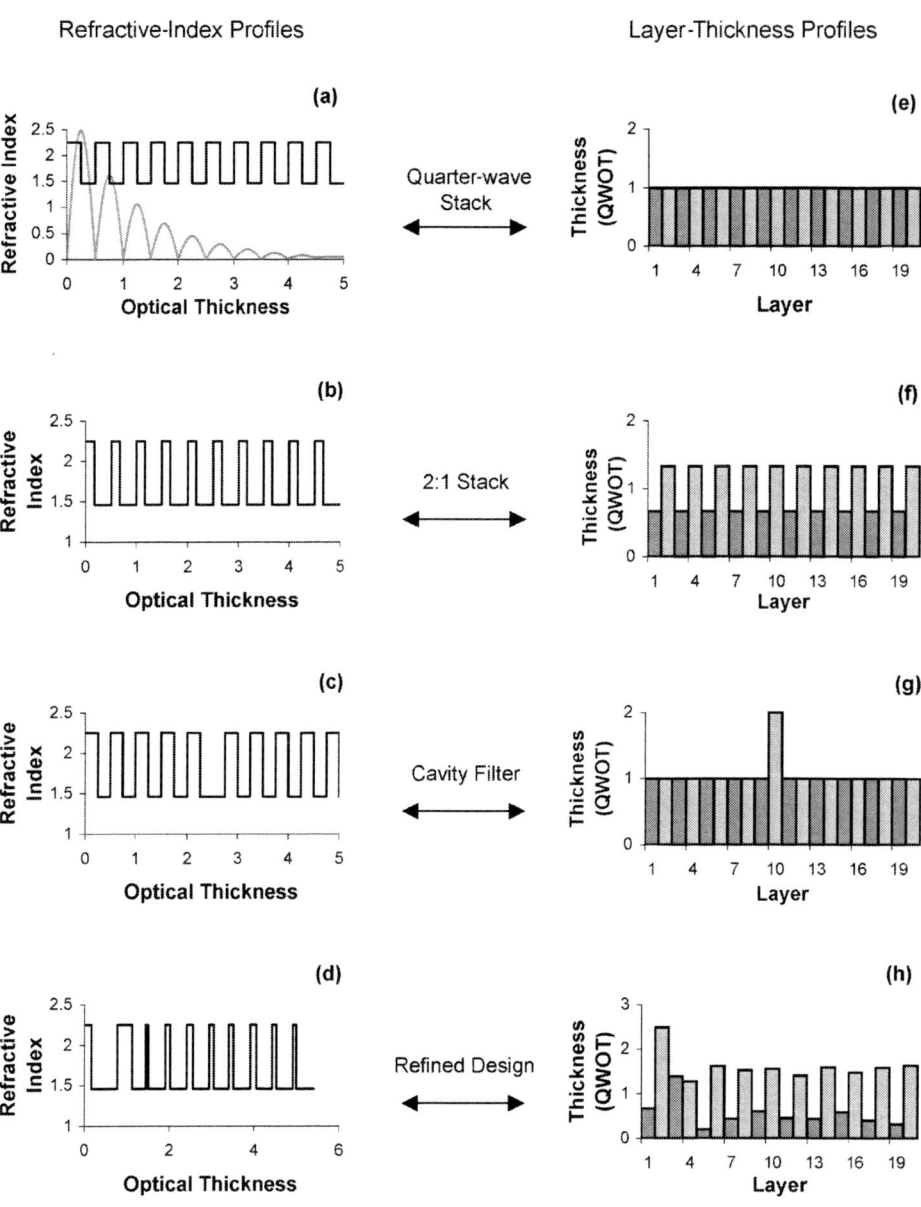

**Figure 1.2 Refractive-index and layer-thickness profiles for four common designs.
(a)–(d): refractive-index profiles as a function of optical thickness (waves); (e)–(h):
corresponding layer-thickness profiles with dark and light shaded bars representing
high and low refractive index. In (a), the parallel electric field is overlayed (arbitrary
units).**

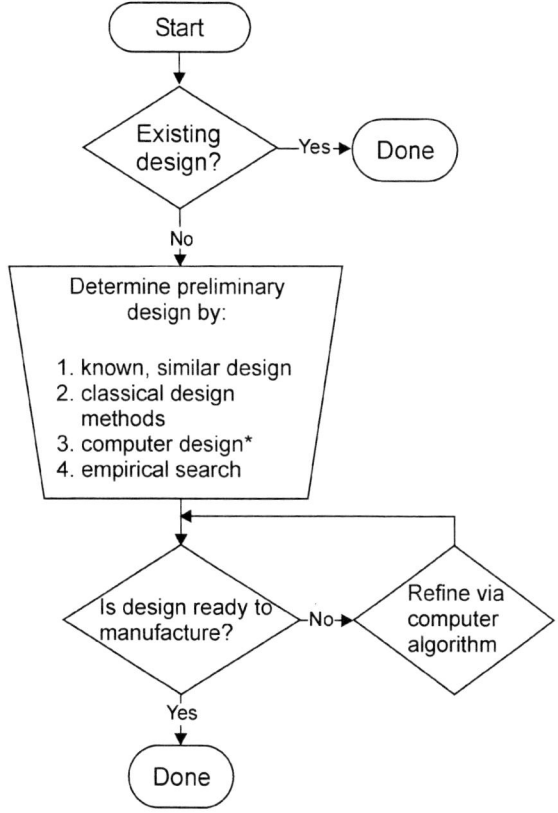

Figure 1.3 Generic flowchart for thin-film design options.
*A computer determines the design from preselected materials.

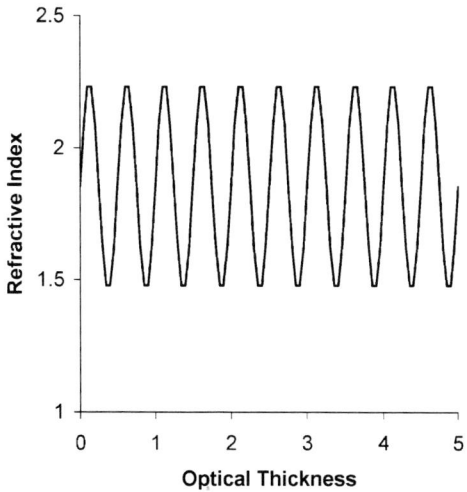

Figure 1.4 Refractive-index profile of a rugate coating.

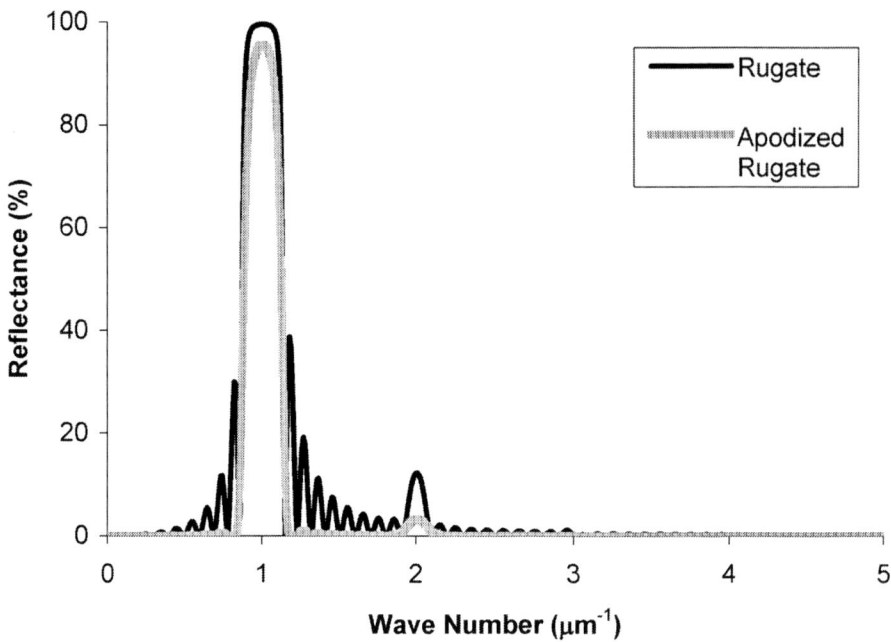

Figure 1.5 Spectral reflectance of rugate and apodized rugate coatings (Figs. 1.4 and 1.6, respectively). Note the supression of high-order stopbands.

Refractive-Index Profile

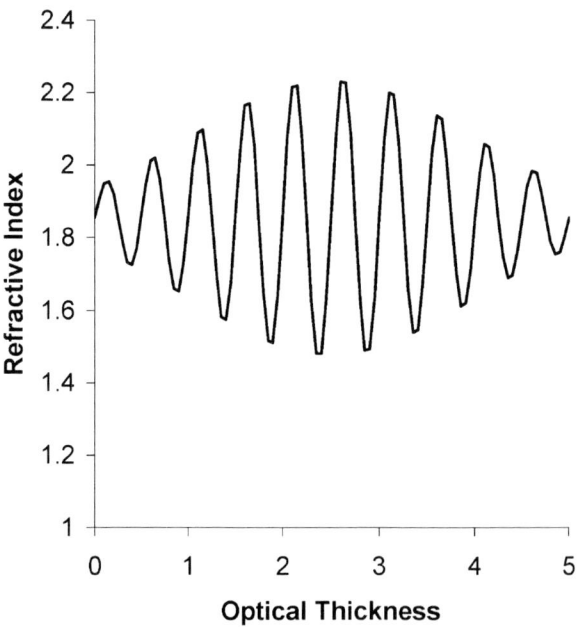

Figure 1.6 Refractive index of a Gaussian-apodized rugate coating.

Sinusoidal Thickness Modulated Designs

Many functions could be employed to create unique patterns of layer thickness in discrete, multilayer thin-film coatings and the corresponding spectral performance. Discrete, layered quarter-wave stacks are routinely used for their stopband properties at harmonics of the center wavelength of the first-order stopband. An inhomogeneous rugate design produces a stopband by sinusoidal modulation of the continuously varying refractive index. This chapter investigates the sinusoidal modulation of the optical thickness of discrete, homogeneous layers in a quarter-wave stack, as well as the synthesis of these thickness-modulated designs with a rugate-equivalent design.

For sinusoidal modulation of the optical thickness of the discrete layers in a quarter-wave stack, the analytical equation used to determine each layer thickness takes the form $T[1 + k\sin(2\pi fL)]$, where T is the QWOT of the quarter-wave stack, k is the modulation amplitude, f is the modulation frequency, and L is the layer number. The sine function was selected first because it is the primary refractive-index modulation function for rugate filters. In contrast to the high-order stopband suppression achieved by rugate filters (see Fig. 1.5 in Chapter 1), sinusoidal modulation of discrete layers produces several stopbands that are harmonics and non-harmonics of the first-order stopband of an unmodulated quarter-wave stack. The purpose of this chapter is to investigate the relationships between the sinusoidal modulation parameters and the resulting spectral performance.

2.1 Review of Stopbands and Passbands

The coating designs covered in this chapter start from the first-order quarter-wave stack of alternating high and low refractive-index layers with equal optical thickness. This type of quarter-wave stack produces stopbands and passbands as shown in Fig. 2.1. The designs presented later are also periodic structures, so it is helpful to briefly review the rigorous mathematical definition for stopbands and passbands.

2.1.1 Periodic thin-film structures

For a group of thin-film layers in a multilayer thin-film design, the corresponding characteristic matrix is readily determined from Eq. (1.10); $\mathbf{P}$ denotes the characteristic matrix of the group of N individual layers:

$$\mathbf{P} = \begin{bmatrix} p_{11} & p_{12} \\ p_{21} & p_{22} \end{bmatrix} = \prod_{i=1}^{N} \mathbf{M}_i. \tag{2.1}$$

In Eq. (2.1), $\mathbf{M}_i$ is the characteristic matrix of the i-th individual layer. If this period or group of layers represented by $\mathbf{P}$ is repeated N times, the resulting characteristic matrix for the entire periodic structure takes the general form

$$\mathbf{P}_1\mathbf{P}_2\mathbf{P}_3\ldots\mathbf{P}_N = \mathbf{P}^N, \tag{2.2}$$

where the subscript of the matrix $\mathbf{P}$ indicates how many times it is repeated.

As explained by Born and Wolf,[2] the matrix $\mathbf{P}$ is unimodular, where its determinant equals unity; and as such, when raised to the N-th power, it can be solved using Chebyshev polynomials of the second kind. Accordingly, $\mathbf{P}^N$ can be solved from Eq. (2.3):

$$\mathbf{P}^N = \begin{bmatrix} p_{11}C_{N-1}(a) - C_{N-2}(a) & p_{12}C_{N-1}(a) \\ p_{21}C_{N-1}(a) & p_{22}C_{N-1}(a) - C_{N-2}(a) \end{bmatrix}. \tag{2.3}$$

$C_i(a)$ represents the Chebyshev polynomials of the i-th order and second kind, where the argument a is given by

$$a = \frac{1}{2}(p_{11} + p_{22}). \tag{2.4}$$

The first seven orders are listed in Appendix B for reference as a function of a. The trace of the periodic matrix $\mathbf{P}$ is given by

$$\text{trace}(\mathbf{P}) = p_{11} + p_{22}. \tag{2.5}$$

The argument a in Eq. (2.4) is simply one-half of the trace. Based on Eqs. (2.4) and (2.5), the argument of the Chebyshev polynomials can be redefined as a function of the trace of $\mathbf{P}$ by variable substitution (see Exercise 2 at the end of this chapter).

2.1.2 Stopbands and passbands

As stated in Ref. [2], when the argument of the Chebyshev polynomials, a, is less than -1, the reflectance of the corresponding periodic matrix $\mathbf{P}$ will increase as N is increased ($\mathbf{P}^N$) for a given wavelength. Otherwise, the reflectance would vary depending on N (Ref. [22]). Furthermore, the wavelength region(s)—with increasing reflectance as N is increased (stopband or high-reflectance band)—is defined, where[23] the stopband is

$$|a| \geq 1. \tag{2.6}$$

Otherwise, a passband exists for wavelengths where Eq. (2.6) is not satisfied.

The edges of a stopband are then defined by

$$|a| = 1. \tag{2.7}$$

Using Eq. (2.7), the spectral half-width of the stopbands for a simple quarter-wave stack is readily determined by Eq. (2.8):[5]

$$\Delta g = \frac{2}{\pi}^{-1} \sin^{-1}\left(\frac{n_H - n_L}{n_H + n_L}\right).$$ (2.8)

Here, the characteristic matrix **P** for a quarter-wave stack is comprised of two layers with equal optical thickness, and Δg is the half bandwidth, or ratio of the center-to-edge wavelengths, of the stopband.

2.2 Sinusoidal Modulation

To allow for full-wave, sinusoidal modulation of the thickness of discrete layers in a multilayer thin film, the average thickness must be greater than or equal to the modulation amplitude. With a given design's average thickness normalized to unity, the thickness of the L-th layer is given by

$$T(L) = 1 + k \sin[\phi(L)],$$ (2.9)

where

$$T(L) = \text{optical thickness of the } L\text{-th layer, } L = 1, 2, 3, \dots N,$$ (2.10a)

$$k = \text{ modulation amplitude, and}$$ (2.10b)

$$\phi(L) = \text{modulation angle.}$$ (2.10c)

To prevent clipping of the full wave, k is constrained by

$$0 \le k \le 1.$$ (2.11)

The modulation angle is further defined as

$$\phi(L) = 2\pi f L + \phi'.$$ (2.12)

Here, f is the modulation frequency and ϕ' is an arbitrary, fixed phase angle. The modulation period is directly defined by

$$T_{\text{base}} = \frac{1}{f}.$$ (2.13)

Integer values of T_{base} correspond to periods of the modulated thin-film design that are an integer number of layers. This will be discussed later in more detail.

The thickness of the discrete layers from Eq. (2.9) can be expressed in terms of quarter-wave optical thickness from Eq. (1.20). Normalizing Eq. (2.9) to the average QWOT, T_{AVG}, gives the general equation used in this chapter for discrete-layer sinusoidal modulation:[24]

$$T(L) = T_{AVG}[1 + k\sin(2\pi fL + \phi')]. \tag{2.14}$$

Equation (2.14) describes the thickness for sinusoidal TMDs. In Eq. (2.14), T_{AVG} is the QWOT of an unmodulated quarter-wave stack. This normalization was chosen so that the modulation amplitude k can vary between 0 and 1 (inclusive) without producing negative layer thickness. Also, a modulation amplitude of 0 produces a design that is equivalent to an unmodulated quarter-wave stack (see Sec. 2.3). Alternatively, Eq. (2.14) can be written as

$$T(L) = T_{AVG}[1 + k\cos(2\pi fL)] \tag{2.15}$$

so the first layer's thickness is close to T_{AVG} when the constant phase term ϕ' is 0. Before looking into the spectral performance of these designs, the range of the modulation frequency f is reviewed.

2.2.1 Modulation frequency range for all possible TMDs

When the constant phase term ϕ' is 0, the range of the modulation frequency that predicts all possible TMDs can be determined. Here, for nonzero, positive integer values of L and f, the modulation frequency range of

$$0 < f_m \le 0.5 \tag{2.16}$$

produces all possible TMDs when $\phi' = 0$. Other TMDs can be produced when $\phi' \ne 0$. These TMD variations are covered in Sec. 2.5.2.

2.3 Degenerate Cases of TMDs (e.g. modulation ampl. k=0)

Certain parameters in Eqs. (2.14) or (2.15) produce TMDs without modulation of the layer thickness, or they will replicate the layer-thickness profiles of high-order quarter-wave stacks. For example, a TMD modulation amplitude of 0 produces a layer-thickness profile that is identical to a first-order quarter-wave stack (all layers have equal optical thickness). For purposes of this text, these types of TMD designs are defined to be degenerate cases of TMDs.

2.3.1 Quarter-wave stacks

A degenerate case of TMDs exists at the upper limit of the modulation frequency range stated in Eq. (2.16), where $f_m = 0.5$. Inserting this modulation frequency into Eq. (2.15) reduces it to

$$T(L) = T_{AVG}[1 + k(-1)^L]. \tag{2.17}$$

Table 2.1 Selected quarter-wave stack designs.

k	Stack	Design
0	Ambient/$(1H1L)^n$/Sub	1:1 quarter-wave stack
1/3	Ambient/$(666H1.333L)^n$/Sub	2:1 quarter-wave stack
1/2	Ambient/$(0.5H1.5L)^n$/Sub	3:1 quarter-wave stack
k	Ambient/$[(1+k)H(1-k)L]^n$/Sub (or Ambient/$[(1-k)H(1+k)L]^n$/Sub)	General form of half-wave layer pairs

From Eq. (2.17), several well-known quarter-wave stacks are described in Table 2.1. The specific case of half-wave layer pairs is described in detail by Thomsen and Wu.[25]

For any quarter-wave stack where the layers repeat every two layers, half of the trace of the corresponding characteristic matrix (a) is determined from Eq. (2.4), where

$$a = \cos(\delta_1)\cos(\delta_2) - \frac{1}{2}\left(\frac{n_H}{n_L} + \frac{n_L}{n_H}\right)\sin(\delta_1)\sin(\delta_2) \qquad (2.18)$$

and

$$\delta_i = \frac{\pi}{2}\frac{\lambda_0}{\lambda}[1 + k(-1)^i] = \frac{\pi}{2}\frac{\sigma}{\sigma_0}[1 + k(-1)^i], \qquad i = 1, 2. \qquad (2.19)$$

In Eq. (2.19), $\delta_{1,2} = \pi/2$ corresponds to the QWOT at the reference wavelength λ_0 or wave number σ_0. Next, when a is plotted versus σ, the spectral positions and the bandwidth of stopbands are determined from Eqs. (2.6) and (2.7). Figures 2.2(a)–(c) show the plots of a versus relative wave number for the first three quarter-wave stacks listed in Table 2.1. In all cases the reference wave number equals 1, corresponding to the first-order stopband. As expected, Fig. 2.2(a) shows a for a 1:1 quarter-wave stack and the stopband positions at relative wave numbers 1, 3, 5, 7, and 9. The stopband positions are also found for the 2:1 and 3:1 quarter-wave stacks, as shown in Figs. 2.2(b) and (c). Using Eq. (2.19), the stopband positions and bandwidth can be found for any value of k.

2.3.2 Progressive series stacks (approximate)

Liddell[26] discusses arithmetic and geometric progressive series stacks. For the arithmetic progression or series stack, the bandwidth of the stopband is broadened beyond that of the first-order, 1:1 quarter-wave stack. Increasing or decreasing the optical thickness of each successive layer in the stack by a common amount accomplishes this (i.e., layer thickness series $T_1, T_1 + d, T_2 + d, \ldots T_n + d$, where T_i is the optical thickness of the i-th layer, d is the common difference in optical thickness for each layer, and $n + 1$ is the total number of layers in the stack). Alternatively, a geometric series stack has the optical thickness of each successive layer scaled by a ratio term r (i.e., $T, Tr, Tr^2, \ldots Tr^n$, where this stack has a total of $n + 1$ layers). The arithmetic series stack (high reflector) is also referred to as a chirped mirror (see Chapter 4).

For the second degenerate case of TMDs, the modulation frequency approaches 0. Here, the base period from Eq. (2.13) becomes very large. For practical designs with some finite number of layers, the degenerate case described here occurs when the following condition is satisfied, or

$$\frac{N}{T_{\text{base}}} < \pi, \tag{2.20}$$

where N is the number of layers in the TMD. A single broadband high-reflectance stopband is produced for selected values of N, T_{base}, and film refractive indices. This degenerate TMD is similar to arithmetic or geometric progression stacks described by Liddell. To illustrate these similarities, the reflectance of a TMD with a low modulation frequency that satisfies Eq. (2.20) is compared with an arithmetic series stack in Fig. 2.3.

2.4 Sinusoidal Thickness Modulated Designs

This section investigates the spectral performance of nondegenerate TMDs and determines where stopbands exist. Within the constraint of Eq. (2.16), modulation frequencies are selected that are small integer values of the corresponding base period from Eq. (2.13). Next, noninteger values of the modulation base period are evaluated to determine the continuity of the spectral performance of TMDs as a function of modulation frequency.

2.4.1 Integer modulation periods

The first TMDs investigated have base periods that are integers from 3 to 10, inclusive (see Table 2.2). Equation (2.13) is used to define the base period as a function of the modulation frequency. Here, the base period is defined as the least integer number of layers that have a unique, nonrepeating thickness pattern. Integer values were selected so the period of the corresponding characteristic matrix is also an integer.

Therefore, at a given wavelength or wave number, Eq. (2.6) determines if a stopband is present. Going back to the layer thickness of these TMDs, Table 2.2 shows the layer thickness for each base period for these TMDs with $k = 0.5$. The modulation amplitude of 0.5 was arbitrarily selected, and other values are discussed later.

Note in Table 2.2 that the number of layers that comprise the periodic structure (shown in the third row) is double the base period T_{base} for odd integers. Here, both the layer thickness and refractive index patterns must repeat in a characteristic matrix of the basic period. The thickness of each layer was determined from Eq. (2.15), where T_{AVG} is set to unity. When the number of layers N

Table 2.2 Layer thickness for selected TMDs for $k = 0.5$.

T_{base}		3	4	5	6	7	8	9	10
f		0.3333	0.25	0.2	0.1667	0.1428	0.125	0.1111	0.1
No. of layers in period		6	4	10	6	14	8	18	10
1	H	0.75	1.00	1.1545	1.25	1.3117	1.3536	1.3830	1.4045
2	L	0.75	0.50	0.5955	0.75	0.8887	1.0000	1.0868	1.1545
3	H	1.50	1.00	0.5955	0.50	0.5495	0.6464	0.7500	0.8455
4	L		1.50	1.1545	0.75	0.5495	0.5000	0.5302	0.5955
5	H			1.5000	1.25	0.8887	0.6464	0.5302	0.5000
6	L				1.50	1.3117	1.0000	0.7500	0.5955
7	H					1.5000	1.3536	1.0868	0.8455
8	L						1.5000	1.3830	1.1545
9	H							1.5000	1.4045
10	L								1.5000

for a given TMD is equal to an integer multiple of the actual period (third row in Table 2.2), the numerical average thickness of the N layers is equal to T_{AVG}.

Example plots of layer-thickness profiles and corresponding plots of one-half of the trace a are shown in Fig. 2.4 for five TMDs from Table 2.2. Figure 2.5 shows the corresponding spectral reflectance for these TMDs with base periods of 3, 4, 5, 8, and 10.

2.4.2 Spacing or position of TMD stopbands

The spectral center of each stopband (A and B) is directly determined from numerical data presented in Figs. 2.5(a)–(e). Here, the ratio of the spectral centering of the two stopbands A and B is shown graphically in Fig. 2.6 for all TMDs listed in Table 2.2. The relationship between this ratio and the modulation frequency is linear and is given analytically by[24]

$$R_{AB} = \frac{\sigma_B}{\sigma_A} = 2f + 1, \quad (2.21)$$

where R_{AB} is the spectral centering ratio of the two adjacent stopbands (A and B) and f is the modulation frequency from Eq. (2.15). To design a TMD that will produce two stopbands with a given spectral spacing, Eq. (2.15) is readily solved for f by

$$f = \frac{1}{2}(R_{AB} - 1). \quad (2.22)$$

Section 2.5 will show that the ratio R_{AB} is independent of the modulation amplitude k and film refractive indices. First, a brief review of noninteger modulation periods will evaluate the continuity of the spectral spacing of adjacent stopbands.

2.4.3 Noninteger modulation periods

Integer values for modulation base periods were evaluated in Sec. 2.4.1 in order for Eq. (2.6) to provide an exact determination of stopband positions. This is also true for many noninteger modulation periods when the following two criteria are met:

1. when the lowest possible integer multiple of the noninteger base period is an integer (I), and
2. when the layer thickness and refractive index patterns repeat every I layers (even number) in the TMD.

For example, a TMD base period of 4.5 can be doubled to be an integer, and then doubled again so that the refractive index pattern or sequence of the layers in the actual period all repeat every 18 layers.

When the period of a TMD does not meet the above two criteria, there are two additional options where Eq. (2.6) determines stopband positions:

1. set the TMD period equal to the total number of layers in a given or proposed TMD $(I = N)$; or
2. approximate by increasing the base period by integer multiples until the result is close to an integer in value.

These two options are discussed in more detail in Sec. 2.5.

Figures 2.7(a) and (b) show the spectral reflectance of two TMDs with noninteger base periods of 4.5 and 8.5, respectively. From direct calculation, the spacing of the two stopbands was determined and is shown in Fig. 2.6. Additional TMDs with noninteger base periods have been evaluated but are not discussed here.

2.5 Determination of All Possible TMD Stopbands

As described in Sec. 2.1.2, Eq. (2.6) determines the spectral position of stopbands. This section evaluates the actual position of stopbands produced by TMDs over a wide spectral range. From these direct calculations, the spectral spacing of these stopbands is determined. Finally, the analytical relationship of the stopband spacing is determined.[27]

For a given nonabsorbing, nondispersive substrate; an ambient medium; two film materials; and a TMD, a FORTRAN program was written that determines the spectral position of all stopbands. The source code for this program is provided in Appendix C. This program accepts several TMDs with different modulation frequencies.

Figures (2.8) to (2.11) show the stopband positions for several TMDs at normal incidence with refractive indices of 1.0, 1.45, 2.25, and 1.52 for the ambient, two films, and substrate media, respectively. The high refractive-index film is adjacent to the incident or ambient medium. The stopbands for each TMD are plotted as horizontal dashes in each figure, where each modulation frequency represents a TMD (vertical axis) and the spectral performance (stopbands) for that TMD is shown over the spectral range of 0.0 to 5.0 wave number (μm^{-1}). The wave number of 1.0 corresponds to the average thickness of each TMD shown. These four figures show the TMD stopband positions for modulation amplitudes of 0.25, 0.5, 0.75, and 1.0, respectively.

In Figs. 2.8 to 2.11, there are several thin line segments that start at $f_m = 0$ with wave numbers 1, 3, and 5 μm^{-1}. These line segments intersect at the center or close to the center of each stopband shown (horizontal segments). For clarity, some of these lines are shown in Fig. 2.12 with the corresponding values of M and N. This group of line segments can be defined analytically by Eq. (2.23), where

$$\sigma_{M,N} = \sigma_0[2Nf_m + (2M - 1)]. \tag{2.23}$$

Here, M and N can be any integer, and $\sigma_{M,N}$ is the spectral position of all possible stopbands produced by a TMD. This universal stopband equation (USE) is independent of modulation amplitude and refractive indices.

The order of the TMD stopband can also be defined by M and N. When $M > 0$ and $N = 0$, the stopband positions for the degenerate case of a quarter-wave stack are determined ($f_m = 0.5$). In this case, these stopbands are odd harmonics of the first-order stopband (as expected). They also exist for some nondegenerate TMDs but depend on modulation frequency. For the general case of TMDs where $N \neq 0$, nonharmonic stopbands are present as shown in Figs. 2.8 to 2.11. Furthermore, TMD stopbands exist at nonharmonic spectral frequencies below the fundamental spectral frequency σ_0, when $M < 0$ and $N < 1$.

In general, the intersection points with the highest number of line segments (determined by USE) occur when the base period of the TMD is an integer. The degenerate TMD case ($f_m = 0.5$) has the lowest number of possible stopbands. Also, as the modulation frequency approaches 0, generally the spacings of stopbands converge and the number of stopbands increases.

For different refractive indices, the plots of the same TMDs in Figs. 2.8 to 2.11 are shown in Appendix D for some common combinations of dielectric materials. Also shown in Appendix D is a table that shows the modulation frequency, base period, and number of layers for each TMD in each set of tables.

2.5.1 Dispersion correction for spectral position of TMD stopbands

This section briefly covers the correction to the USE for TMDs [Eq. (2.23)] for dispersion in the refractive indices of the two film materials n_L and n_H. In general, the total optical thickness of the low and high index layers in a TMD are equal. This is true for TMDs with integer base periods and approximately true for noninteger base periods. Therefore, Eq. (2.23) can be rewritten to correct for dispersion at normal incidence, where

$$\sigma_{M,N} \simeq \frac{\sigma_0[2Nf_m + (2M - 1)]}{1 + \dfrac{1}{2}\left(\dfrac{\Delta n_L}{n_{L_0}} + \dfrac{\Delta n_H}{n_{H_0}}\right)}. \tag{2.24}$$

Here, Δn_i represents the difference in the two films' refractive indices at wavelengths that correspond to the TMD stopbands of order (1,0) and (M, N); n_{i0} represents the film refractive indices at the fundamental stopband position (or wavelength) σ_0. For example, a TMD with parameters $k = 0.5$, $f = 0.25$, $L = 40$, $\lambda_0 = 1$ μm, and nondispersive refractive indices of 1.46, 2.25, and 1.52 produces a fundamental TMD stopband of order (1,0) at $\lambda_0 = 1$ μm and a second desired stopband of order (1,1) at $\lambda = 0.667$ μm, as shown in Fig. 2.13. If the two films' refractive indices were 1.47 and 2.28 near the wavelength of 0.667 μm, Eq. (2.18) could correct for this dispersion and predict the spectral center of the (1,1)-order TMD stopband at $\lambda = 0.673$ μm. Figure 2.13 also shows the spectral performance of the same TMD as described above except with the stated dispersion. The derivation of the dispersion factor in Eq. (2.18) is included as Exercise 10 at the end of this chapter.

2.5.2 Effect of a nonzero phase term on TMD stopband positions

As mentioned in Sec. 2.5.1, the total optical thickness of all low and high refractive index layers is generally equal. This is true when the given TMD has an integer base period and the total number of layers is an integer multiple of the number of layers in the base period. For all TMDs and analysis mentioned so far, the constant phase term ϕ' from Eq. (2.14) has been set equal to 0. This section investigates the spectral centering of TMD stopbands for one TMD with different values of ϕ'.

Figure 2.14 shows the reflectance of several TMDs with all of the design parameters constant except for ϕ'. In this figure, the spectral position of the TMD stopbands is slightly changed. The stopbands with the three lowest orders show minimal change in spectral position or reflectance. However, the reflectance of the higher-order TMD stopbands is significantly changed. Also note that the ripple structure in the passbands is affected by ϕ'. Other TMD parameters produce similar results.

2.6 Electric Field Analysis of a TMD

Other researchers have reported on the analysis of electric fields in multilayer coatings and methods to reduce electric fields and corresponding laser damage.[28–30] The parallel electric field as a function of optical distance from the incident medium of an arbitrary TMD is shown in Figs. 2.15(a) and (b). Here, an optical distance of 0.25 is equal to one quarter-wave for a given wavelength. In these figures, the calculated electric fields are at the wavelengths at the center of the (1,0)- and (1,1)-order TMD stopbands. The quasi-exponential decay envelope of the field strength is similar for both stopbands. For the general case of a double quarter-wave stack design (to highly reflect two wavelengths), the electric field amplitude remains high through the outer stack (see Fig. 2.16). However, this problem is solved for high-order stacks when the second wavelength is a harmonic of the first; TMDs also overcome this problem for harmonic and nonharmonic wavelengths.

2.7 Reflectance Phase Shift of a TMD

At the center wavelength of the first-order stopband of a quarter-wave stack, the differential phase shift is equal to 0 according to the Abelès phase convention. The differential phase shift in reflection is shown for a quarter-wave stack in Fig. 2.17(a). Note that the phase shift is relatively smooth through the wavelength region of the first-order stopband. Abelès determined an analytical expression for the slope of the absolute phase shift at the center wavelength of the stopband at normal incidence.[31] For nondispersive materials, this expression is given by

$$\frac{d\delta}{d\sigma} \cong \frac{\pi n_0}{\sigma(n_L - n_H)},$$
(2.25)

where the absolute phase shift upon reflection is δ. Note that the slope is negative as a function or wave number and positive for wavelength. Using the effective indices for n_0, n_L, and n_H from Eqs. (1.6) and (1.7), Eq. (2.25) can be rewritten for oblique incidence angles for the p- and s-polarizations.[32] From these expressions, the slope of the differential phase shift Δ can be determined for oblique incidence angles.

Figure 2.17(b) shows the differential phase shift in reflection for the TMD used for Fig. 2.13 except the design was tuned for a 45-deg incidence using effective indices. In this case, the slope of the differential phase shift is larger than for a quarter-wave stack with the same film refractive indices.

2.8 Applications

2.8.1 Dual-band high reflector

Assume there is a request to design a normal-incidence, high-reflectance coating for the approximate laser wavelengths of 1064 nm and 633 nm. As previously mentioned, a dual quarter-wave stack can readily accomplish this task. However, if other design constraints such as damage threshold requirements or film stress limits (mechanical) exist, a TMD may provide an advantage. Using Eqs. (2.15) and (2.16), the modulation frequency for a TMD is calculated to be 0.34. Here, the two TMD stopbands are of order (1,0) for 1064 nm and (1,1) for 633 nm. Next, assume the selected film materials have refractive indices of 1.45 and 2.25. Upon inspection of Figs. 2.8 to 2.11, the (1,0)/1064-nm and (1,1)/633-nm TMD stopbands are produced for the modulation frequency of 0.34 and at modulation amplitudes lower than 0.75. The reflectance at 1064 nm decreases with increased modulation amplitude, and reflectance at 633 nm reaches a maximum near the modulation amplitude of 0.7. These results are shown in Fig. 2.18. Here, the modulation amplitude of 0.625 could be selected for highest reflectance at both wavelengths.

2.8.2 Triple-band high reflector

Three wavelengths were to be highly reflected at normal incidence: 1064 nm, 640 nm, and 514 nm. Using Eq. (2.21), the ratios of the spectral centering of the two stopbands (640 nm and 514 nm) to the longer-wavelength stopband were calculated to be 1.6625 and 2.07, respectively. Figure 2.9 was evaluated for the presence of stopbands near the same wave numbers as these ratios, and the refractive indices of the film materials were chosen to be the same as in Fig. 2.9. Stopbands were present for these ratios near the modulation frequency of 0.125. This trial modulation frequency, $f_m = 0.125$, was inserted into Eq. (2.23) to determine if two integer pairs M and N exist that determine the same spectral centering ratio of these stopbands. After testing several values of f_m, M, and N in Eq. (2.23), a close match was obtained for the two ratios, as shown in Table 2.3 when $f_m = 0.107$ and when the values of M and N are as shown. Note that the trial value of f_m must be the same for both ratio calculations to determine one TMD.

Next, the number of layers for this TMD was selected to be 64. Then the modulation amplitude k of 0.5 was evaluated to determine if the reflectance was near unity for all three wavelengths. However, the reflectance at 1064 nm, 640 nm, and 514 nm was found to be 71.7%, 99.99%, and 99.97%, respectively. Other values of k were tested until the reflectance at each wavelength was near unity ($k = 0.45$). Figure 2.19 shows the reflectance of the final design. The reflectance of the final design at 1064 nm, 640 nm, and 514 nm was found to be 99.76%, 99.99%, and 99.56%. Further improvements could be made to this TMD by adjusting the modulation frequency to shift the spectral centering of the stopbands.

As alternatives to this procedure, the program from Appendix C can be used to determine if the stopbands exist for the given modulation frequency and other given design input parameters. Also, graphs from Appendix D can be examined to determine if stopbands exist for the precalculated ratios of the center wavelength of each stopband for films with different refractive indices. This direct (graphical) method of designing a TMD is very useful for many TMD applications.

Table 2.3 Test of trial values of f_m, M, and N.

Stopband Ratio (Design Target)	Calculated Ratio	Trial f_m	Trial M	Trial N
$1064/640 = 1.6625$	1.642	0.107	1	3
$1064/514 = 2.0700$	2.07	0.107	1	5

2.8.3 Rugate TMD

This section first considers a homogeneous, discrete-layered TMD with the following parameters: $L = 19$ with the first and last layers having high refractive index, $n_L = 1.46$, $n_H = 2.25$, $n_0 = 1.0$, $n_s = 1.52$, normal incidence, $\sigma_0 = 1.0 \ \mu m^{-1}$, $\sigma_1 = 1.5 \ \mu m^{-1}$, $f = 0.25$, and $k = 0.5$. The refractive-index profile and spectral transmittance are shown in Figs. 2.20 and 2.21, respectively.

Next, to reproduce the spectral performance of the discrete TMD in this example with a rugate TMD, a Q-function was selected to synthesize an inhomogeneous refractive-index profile $n(x)$ (see Sec. 1.4.2). Dr. Pierre Verly at the National Research Council Canada provided the following methods, designs, and data.

The Q-function, used to synthesize a rugate-TMD equivalent of the discrete-layered TMD, was determined by taking the inverse Fourier transform of the discrete-layer index profile in Fig. 2.20. The result was a numerical Q-function (complex).[33]

Finally, the inhomogeneous refractive-index profile of the rugate TMD was synthesized by taking the Fourier transform of the Q-function. Three different rugate TMDs were obtained by computing the Fourier transforms over the wave number regions: $\sigma = 0$–$1.25 \ \mu m^{-1}$, $\sigma = 0$–$2.0 \ \mu m^{-1}$, and $\sigma = 0$–$4.0 \ \mu m^{-1}$. Figures 2.22 to 2.24 show their refractive-index profiles and spectral transmittance. All three rugate TMDs have nearly identical transmittances as the discrete-layer TMD below the maximum wave number of the three spectral regions used in the calculation. Above the maximum, the rugate TMDs have near-unity transmittance. The rugate TMD's index profile more closely matches the discrete-layer TMD as the maximum wave number used for the Fourier transform is increased.

One significant result of this exercise, in agreement with Ref. [33], is that rugate TMDs with high-reflectance stopbands were readily determined using the numerical Q-function; although, in theory, the reflectance should be small for optical thin-film synthesis by the Fourier transform method.

2.9 Exercises

1. Derive the reflectance as a function of the two refractive indices of the layers for the general case of one period of a TMD with $f = 0.5$.
2. Rewrite the Chebyshev polynomials using the trace of the characteristic matrix [Eq. (2.5)] instead of Eq. (2.4) (also see Appendix B).
3. Show that for the center wavelength of a stopband, the trace of the characteristic matrix is greater than 2.
4. Derive Eq. (2.8) from Eq. (2.7) for a quarter-wave stack with its characteristic matrix having two layers of equal optical thickness.
5. Write a software program, or modify the program from Exercise 5 in Chapter 1, to use Chebyshev polynomials to calculate the reflectance of periodic thin-film designs with up to 10 periods, given the characteristic matrix for the layers within the period from Eq. (2.1).

6. For a quarter-wave stack, using Eq. (1.10), determine the general analytical expression for the reflectance of an N-period stack with alternating layers of low and high refractive index at normal incidence.

7. Determine the approximate bandwidth of the first-order stopband of an arithmetic series stack as a function of the film's refractive indices, the common difference in layer thickness, and the number of layers.

8. Write a software program based on Eqs. (2.17) and (2.18) that calculates a as a function of wave numbers 0–10 for $k = 0$, 0.25, and 0.75.

9. Show that the TMD Eq. (2.14) can be rewritten

$$T(L) = (T_{AVG} - k) + \left\{ 2k \sin^2 \left[2\pi \left(\frac{f}{2} \right) L \right] \right\}$$

when the constant phase term in Eq. (2.14), ϕ', equals $-\pi/2$. Note that the TMD modulation frequency can also be defined as shown in the above equation.

10. Derive the dispersion correction factor in Eq. (2.24).

11. Derive a general expression for the phase-shift derivative for nondispersive TMDs with $T_{\text{base}} = 2$ and 4 for a 45-deg angle of incidence (see Ref. [31]).

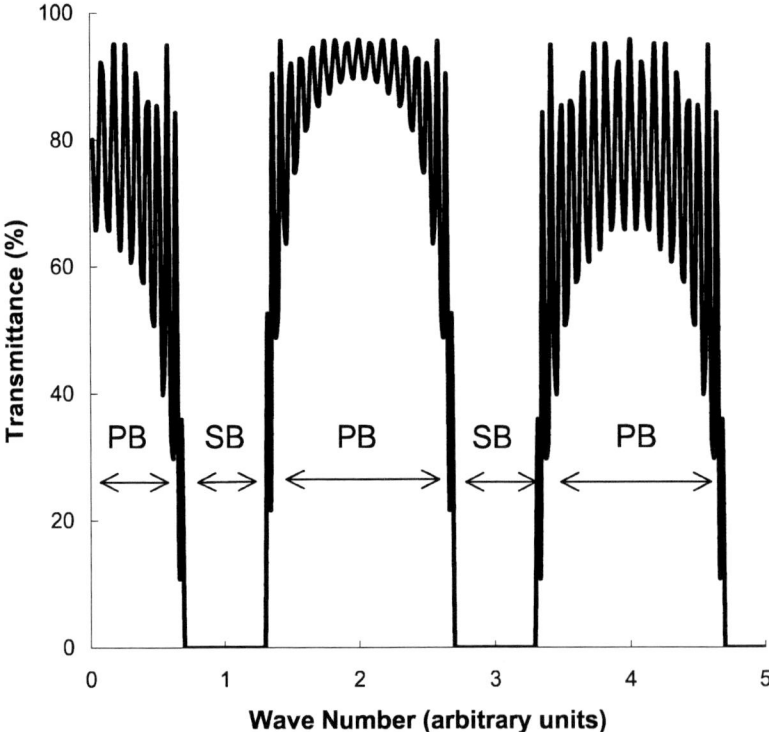

Figure 2.1 Transmittance of an arbitrary quarter-wave stack (shortwave pass filter). Some of the passbands (PB) and stopbands (SB) are indicated in the figure.

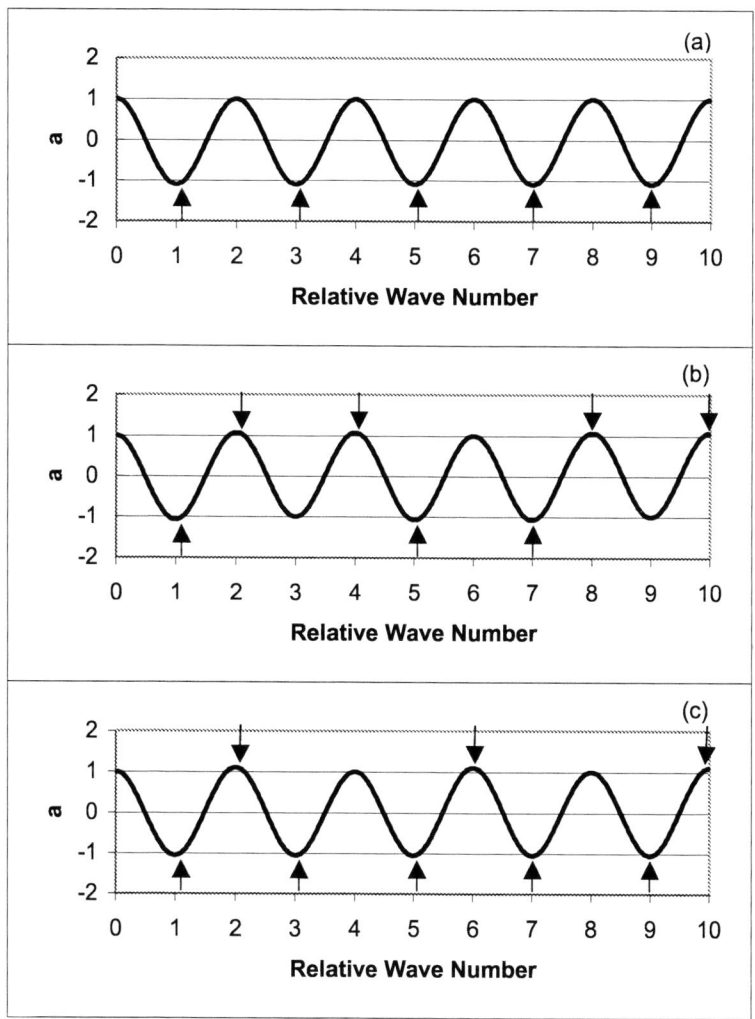

Figure 2.2 The value of a, or one-half of the trace of the characteristic matrix, for the first three quarter-wave stacks listed in Table 2.1. The arrows on (a)–(c) point to the spectral positions of the stopbands.

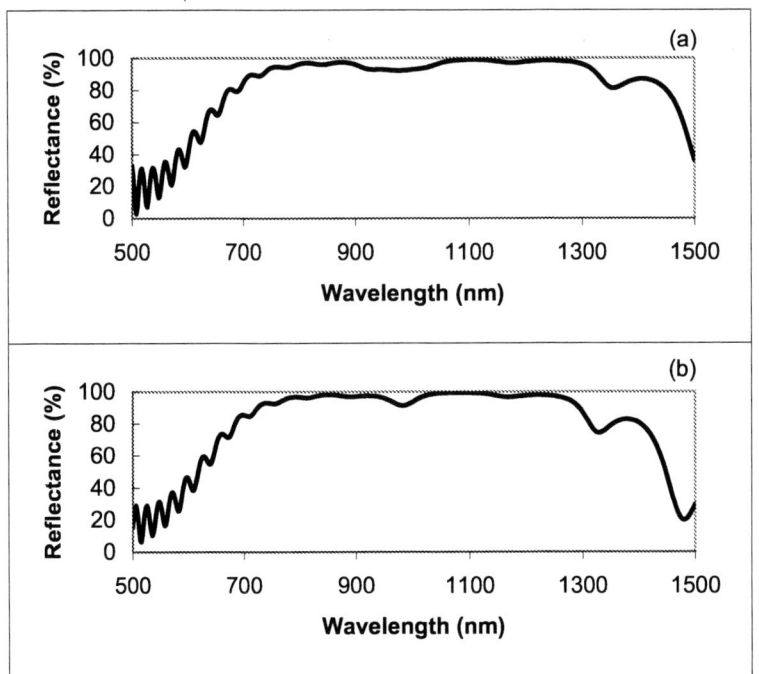

Figure 2.3 Reflectance comparison of a low modulation frequency TMD and a progressive series stack: (a) $f = 0.01$, $k = 0.5$, 26 layers, modulation phase = 225 deg;
(b) first (outer) layer = 669.2 nm QWOT, layer increment = 26.7692 nm QWOT, 26 layers. For both designs, the refractive indices for the ambient, substrate, and two films
are 1.0, 1.52, 1.46, and 2.25, respectively.

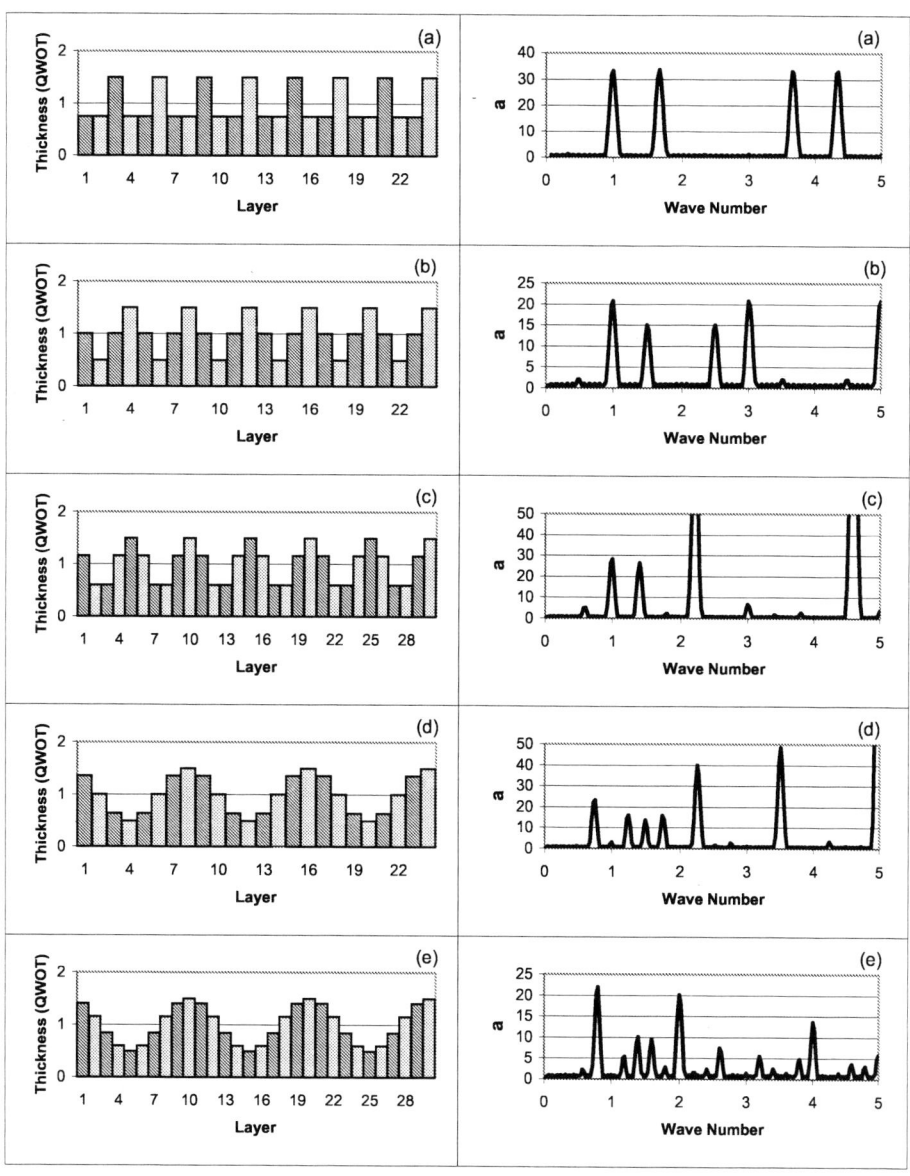

Figure 2.4 Layer-thickness profiles and one-half of the trace for five TMDs described in Table 2.2. Figures (a)–(e) correspond to TMDs with base periods of 3, 4, 5, 8, and 10, respectively. For all designs, the refractive indices for the ambient, substrate, and two films are 1.0, 1.52, 1.46, and 2.25, respectively.

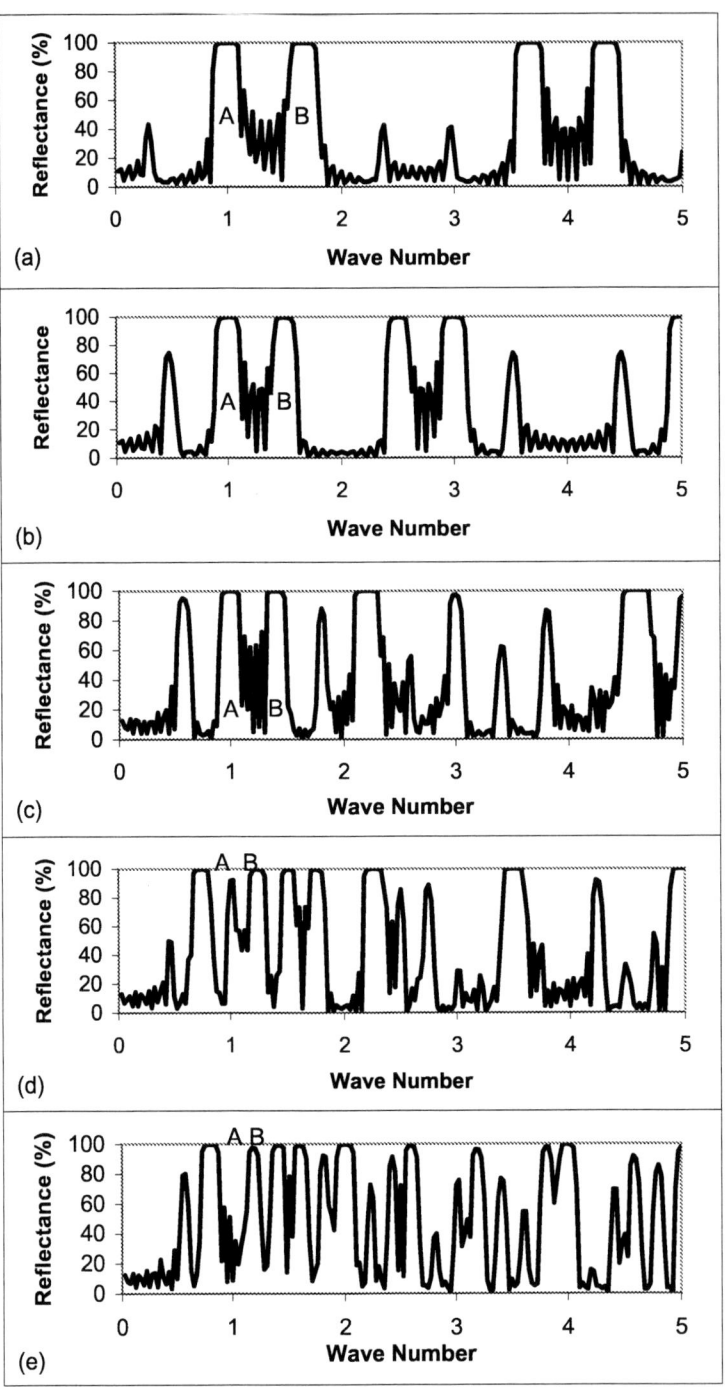

Figure 2.5 Spectral reflectance for five TMDs described in Table 2.2 and Fig. 2.4; (a)–(e) correspond to TMDs with base periods of 3, 4, 5, 8, and 10, respectively.

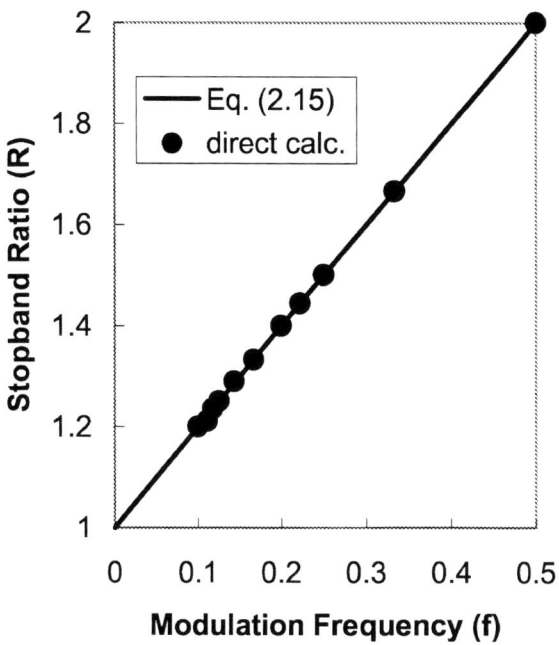

Figure 2.6 Ratio of spectral centering of stopbands A and B from Fig. 2.5 and TMDs in Table 2.2. The line is determined by Eq. (2.15).

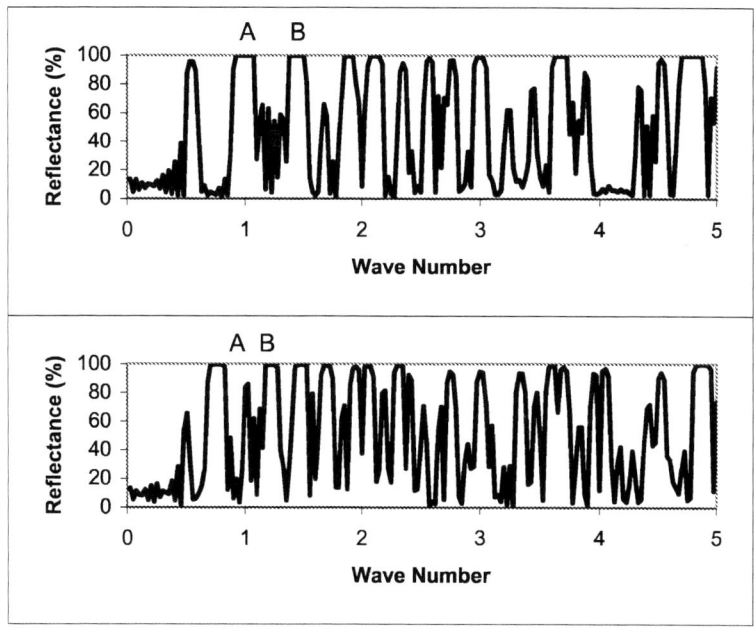

Figure 2.7 Spectral reflectance for two TMDs with base periods of (a) 4.5, and (b) 8.5, respectively. For both designs, the refractive indices for the ambient, substrate, and two films are 1.0, 1.52, 1.46, and 2.25, respectively.

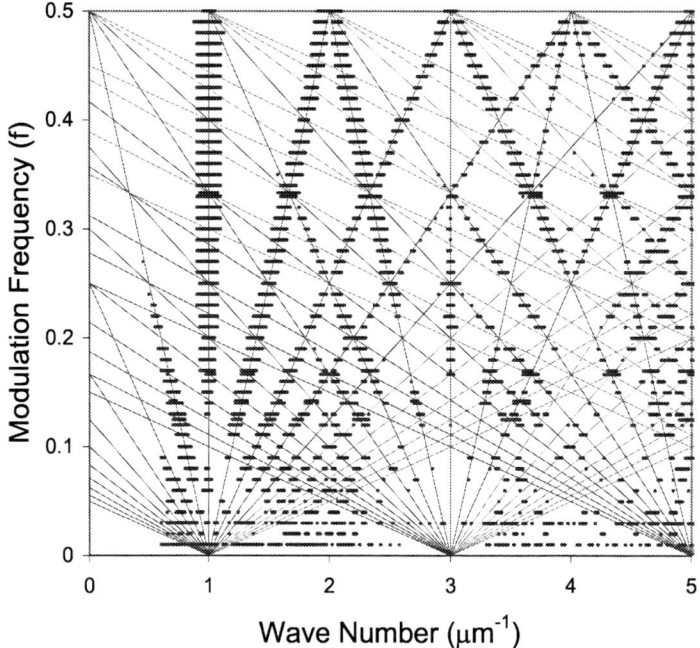

Figure 2.8 Spectral positions of all stopbands for a modulation amplitude of 0.25. For all designs, the refractive indices for the ambient, substrate, and two films are 1.0, 1.52, 1.45, and 2.25, respectively.

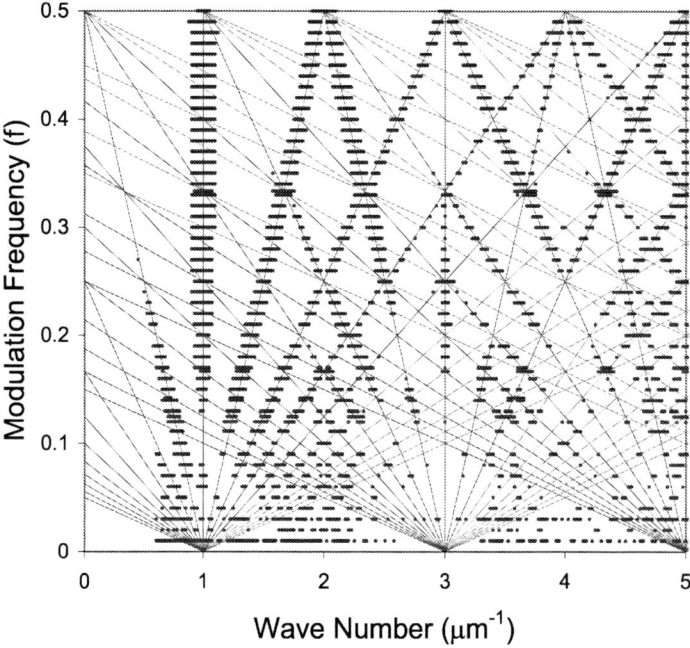

Figure 2.9 Spectral positions of all stopbands for a modulation amplitude of 0.5. For all designs, the refractive indices for the ambient, substrate, and two films are 1.0, 1.52, 1.45, and 2.25, respectively.

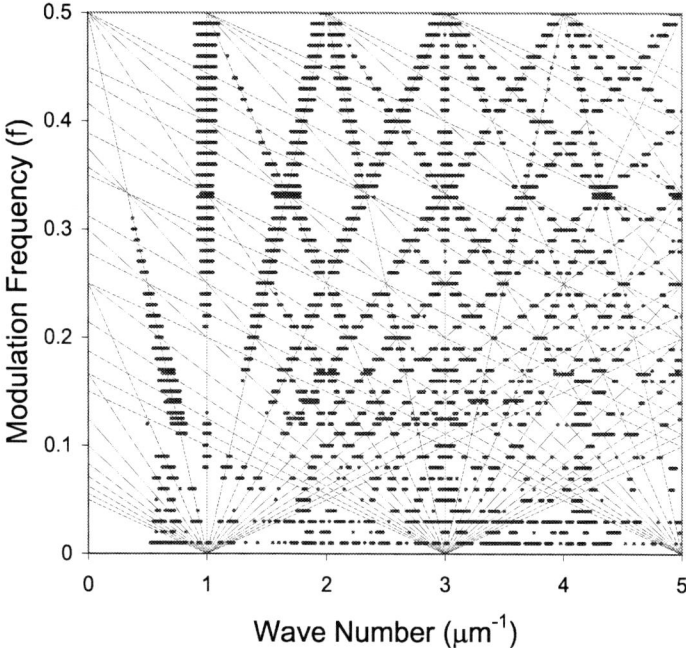

Figure 2.10 Spectral positions of all stopbands for a modulation amplitude of 0.75. For all designs, the refractive indices for the ambient, substrate, and two films are 1.0, 1.52, 1.45, and 2.25, respectively.

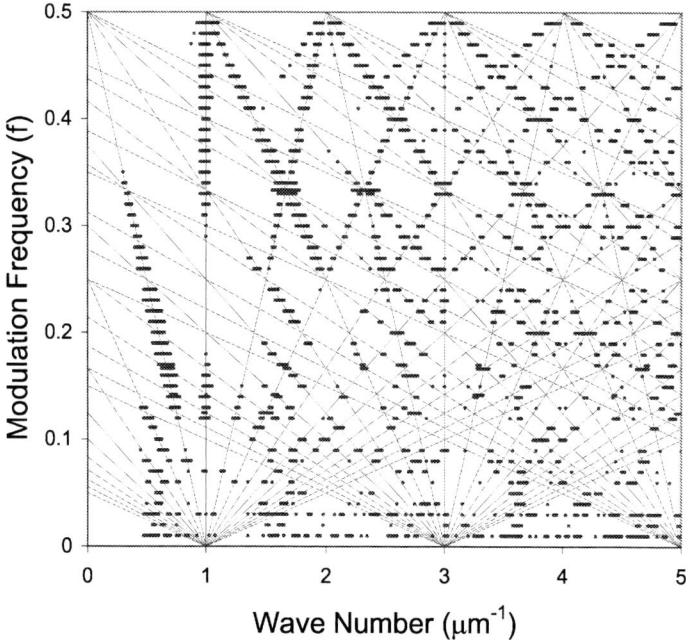

Figure 2.11 Spectral positions of all stopbands for a modulation amplitude of 1.0. For all designs, the refractive indices for the ambient, substrate, and two films are 1.0, 1.52, 1.45, and 2.25, respectively.

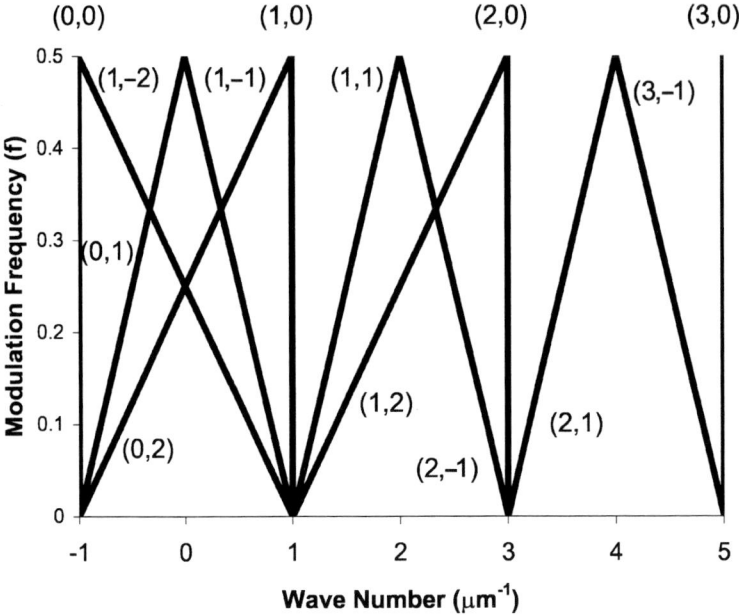

Figure 2.12 Selected line segments as defined by Eq. (2.17). Here, the values of M and N are shown for the displayed lines as (M, N). The vertical segments are identified above the graph.

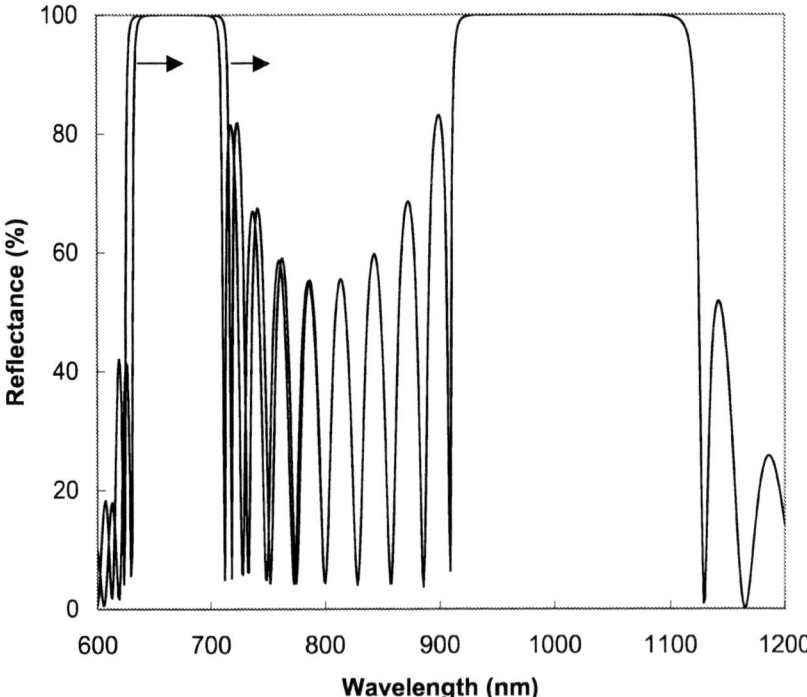

Figure 2.13 Spectral reflectance of a TMD with and without refractive-index dispersion. Equation (2.18) predicts the spectral center of the (1,1)-order TMD stopband at 673 nm.

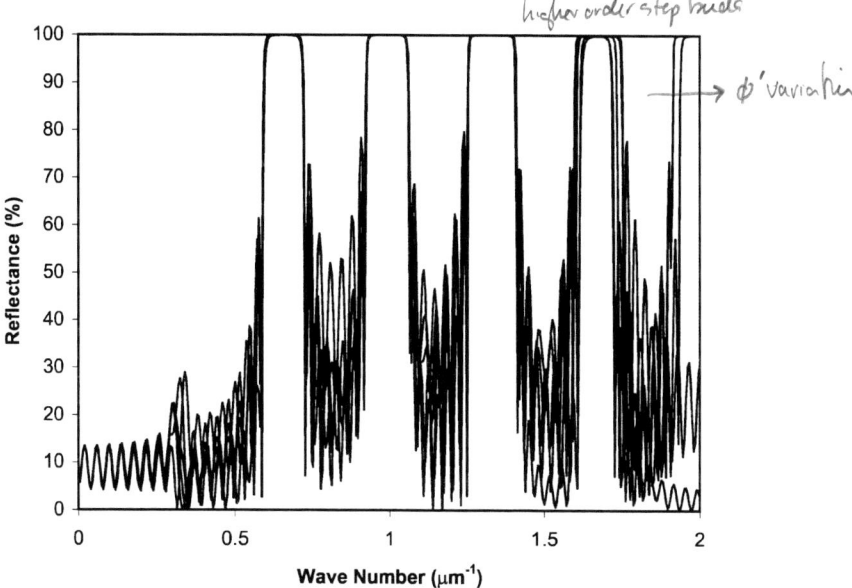

Figure 2.14 An overlay of reflectances for several TMDs, where all have the same parameters except the constant phase angle with values of 0, 30, 45, 90, and 180 deg. The TMD parameters are $f = 0.16667$, $k = 0.5$, $L = 50$, and film refractive indices = 1.46 and 2.25.

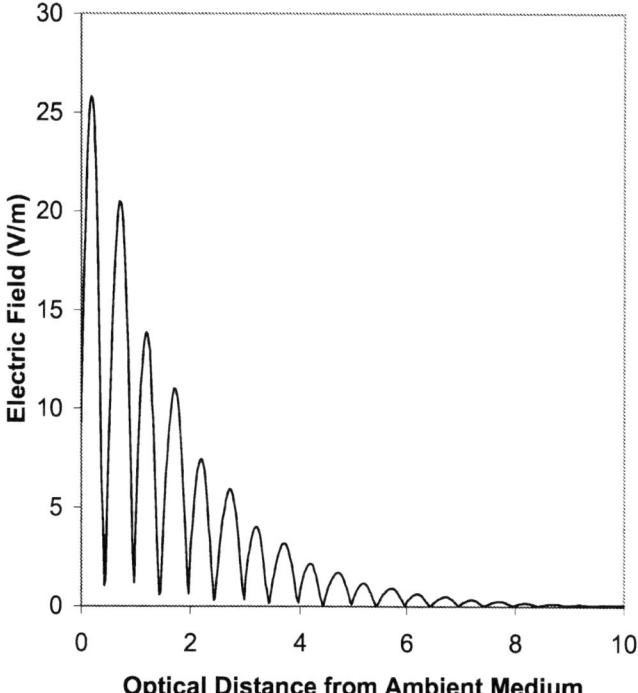

Figure 2.15(a) Electric field for the center wavelength (1000 nm) for an arbitrary TMD with (1,0) stopband. The TMD parameters are $f = 0.25$, $k = 0.5$, $L = 40$ and film refractive indices = 1.46 and 2.25.

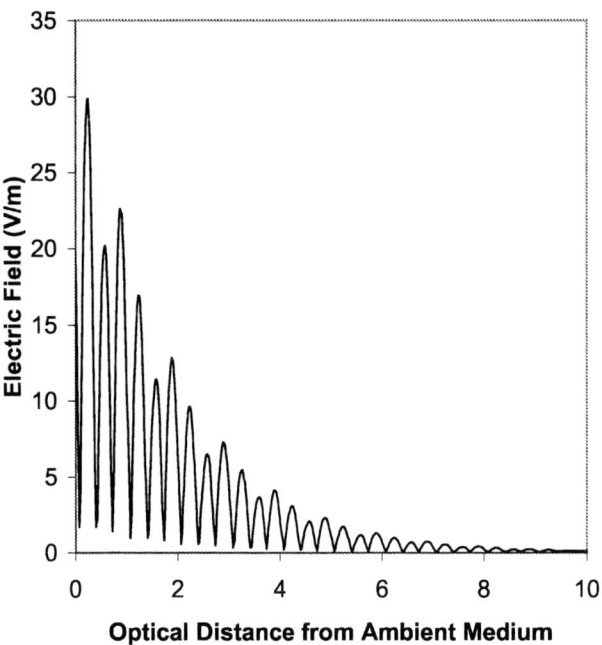

Figure 2.15(b) Electric field for the center wavelength (667 nm) of the same TMD from Fig. 2.15(a) but with a (1,1) stopband.

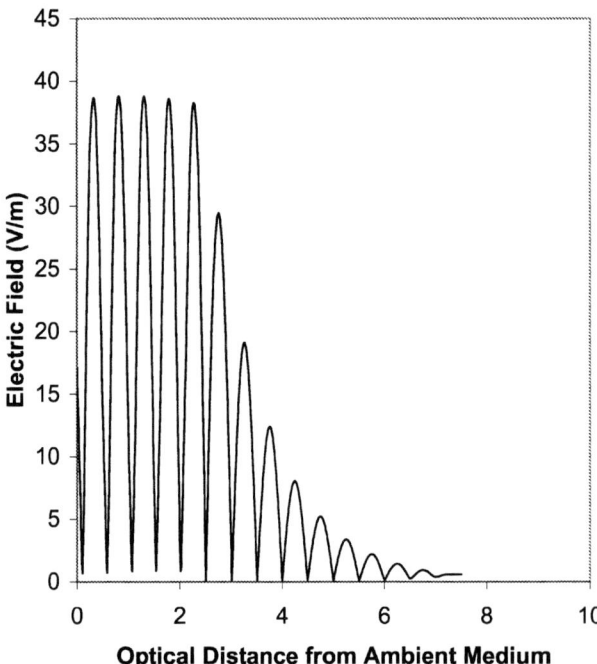

Figure 2.16 Electric field for the stopband center wavelength (1000 nm) for a double quarter-wave stack with a 500-nm QWOT stack adjacent to the ambient medium. The stack parameters are sub/(LH)^{10} (0.5L 0.5H)^{10}/ambient; the film refractive indices = 1.46 and 2.25.

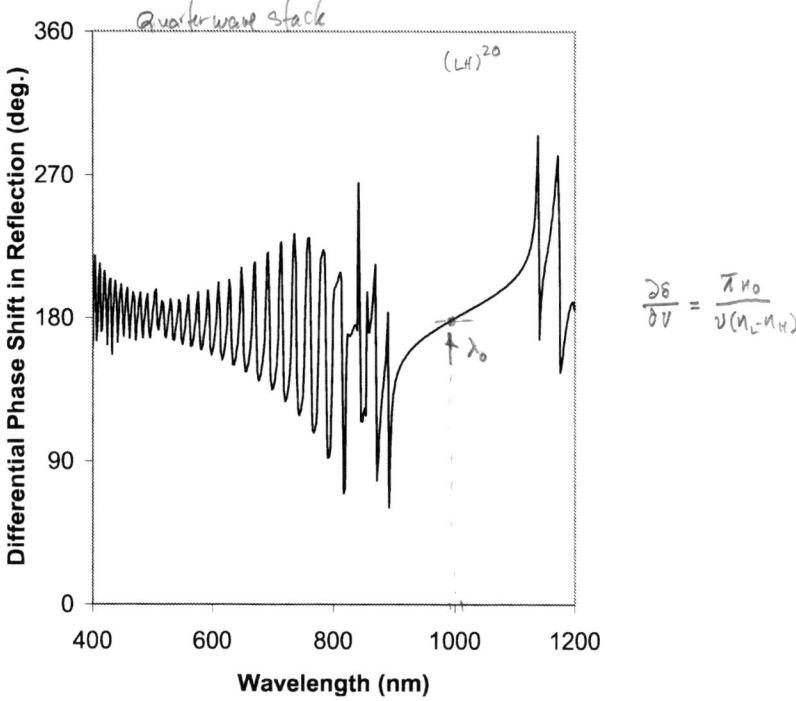

Figure 2.17(a) Differential phase shift in reflection for a quarter-wave stack [(LH)[20]] tuned to a 45-deg angle of incidence. The layers are all QWOT at 1000 nm.

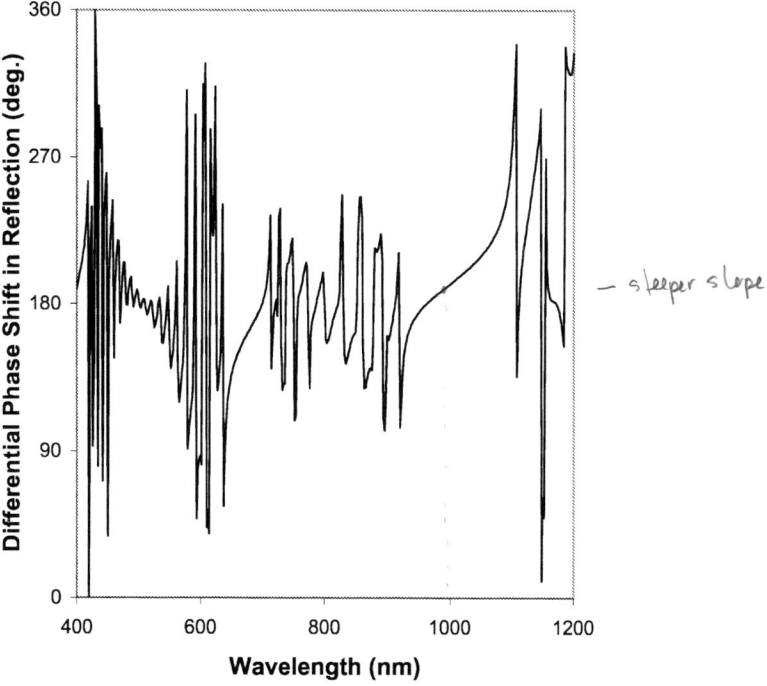

Figure 2.17(b) Differential phase shift in reflection for the TMD from Fig. 2.13 with layer thicknesses tuned to a 45-deg angle of incidence.

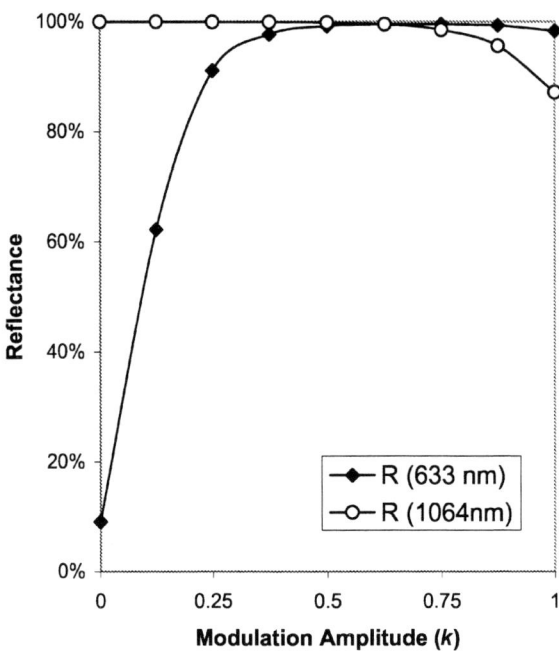

Figure 2.18 The results of 1064-nm and 633-nm reflectance versus modulation amplitude for a 20-layer TMD.

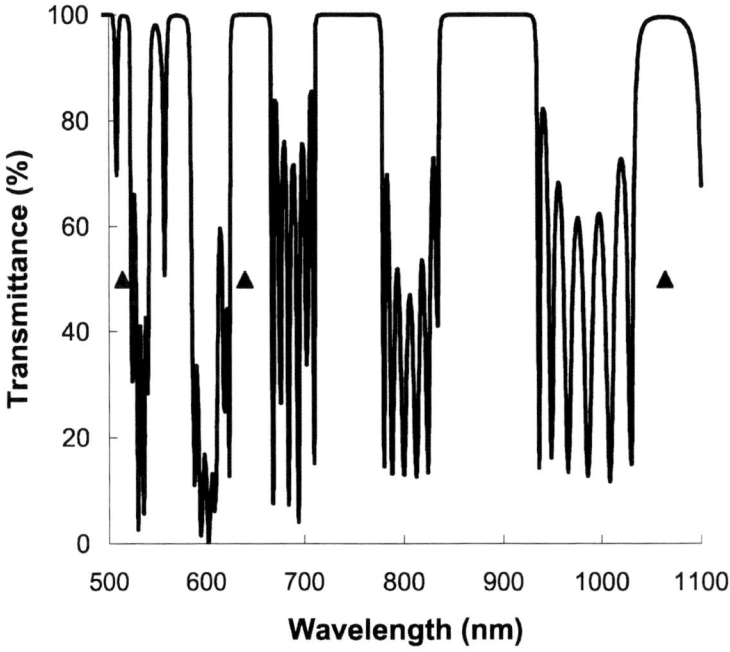

Figure 2.19 Reflectance of the triple-band TMD in Sec. 2.8.2. The TMD parameters are $f = 0.107$, $k = 0.45$, $L = 64$, and film refractive indices = 1.45 and 2.25, normal incidence. Arrows are placed at the three design wavelengths: 514 nm, 640 nm, and 1064 nm.

Refractive-Index Profile

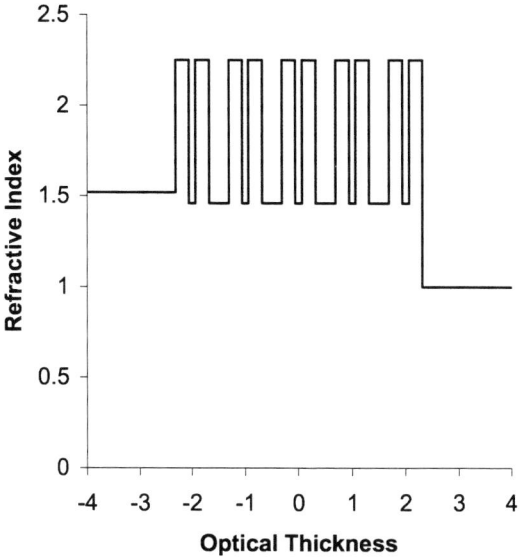

Figure 2.20 Refractive-index profile of the discrete-layered TMD in Sec. 2.8.3.

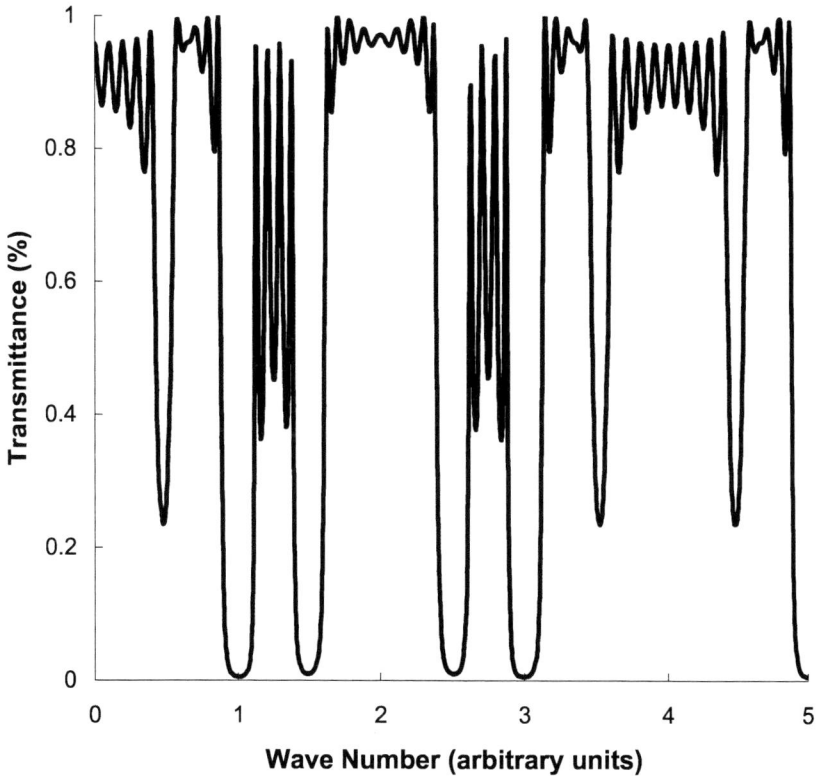

Figure 2.21 Transmittance of the discrete-layered TMD in Sec. 2.8.3.

Refractive-Index Profile

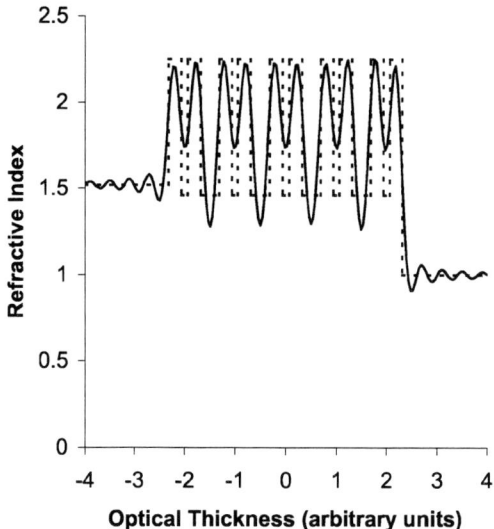

Figure 2.22(a) Refractive-index profile of rugate TMD (solid) and discrete-layered TMD (dashed).

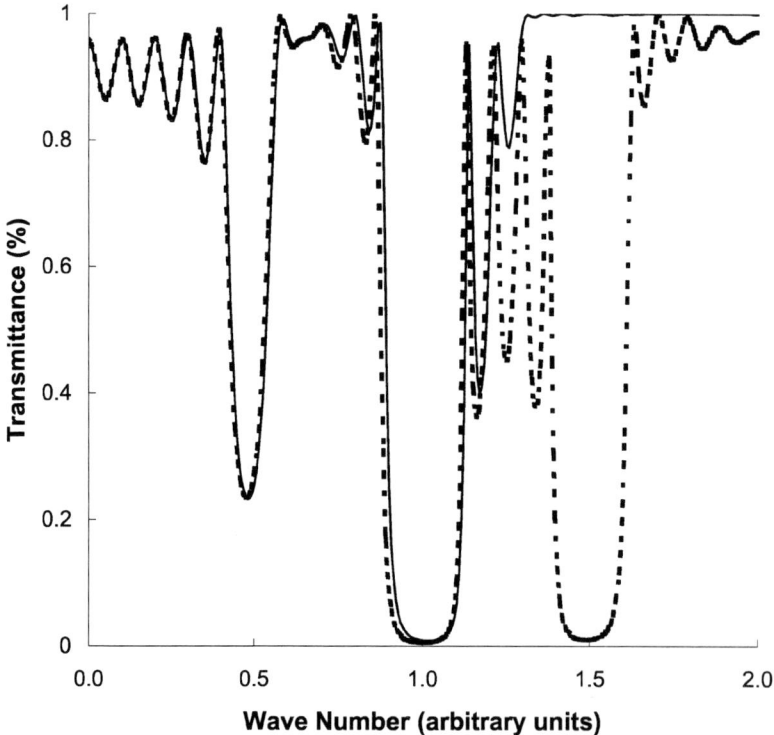

Figure 2.22(b) Transmittance of rugate TMD (solid) and discrete-layered TMD (dashed) in Sec. 2.8.3.

Refractive-Index Profile

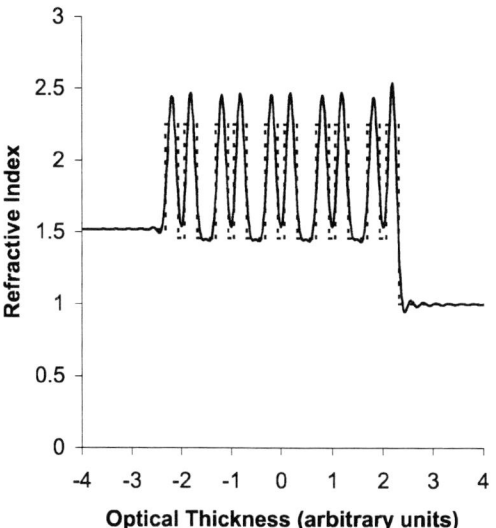

Figure 2.23(a) Refractive-index profile of rugate TMD (solid) and discrete-layered TMD (dashed).

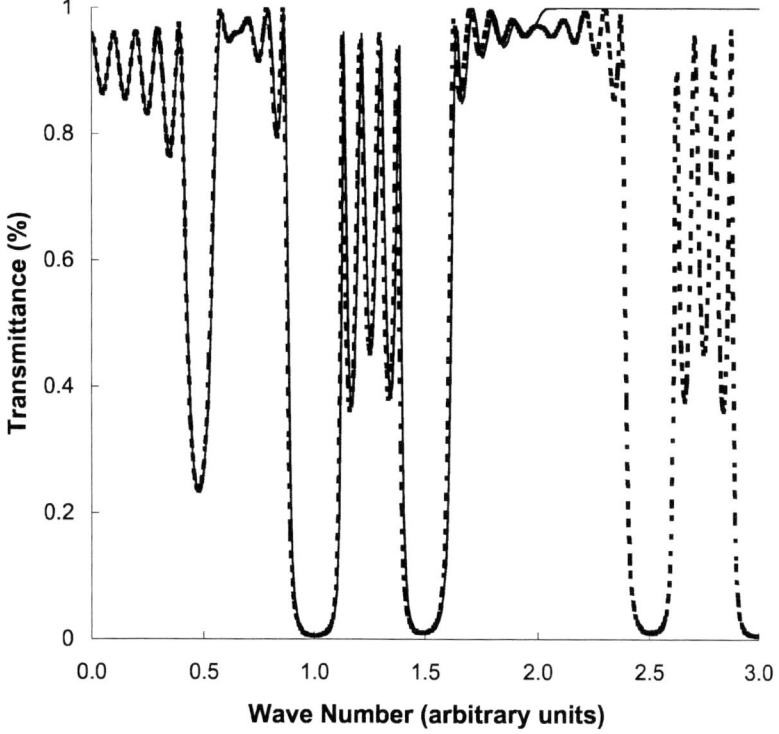

Figure 2.23(b) Transmittance of rugate TMD (solid) and discrete-layered TMD (dashed) in Sec. 2.8.3.

Refractive-Index Profile

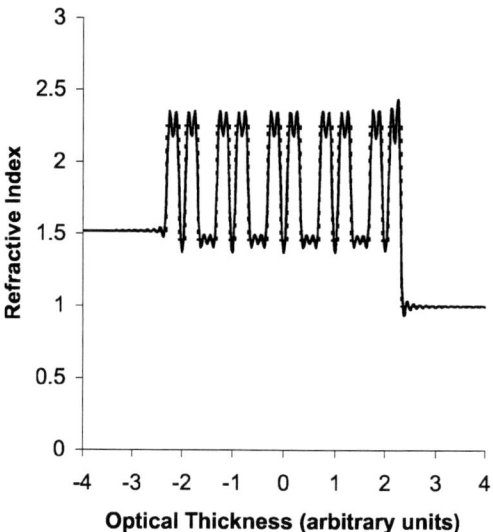

Figure 2.24(a) Refractive-index profile of rugate TMD (solid) and discrete-layered TMD (dashed).

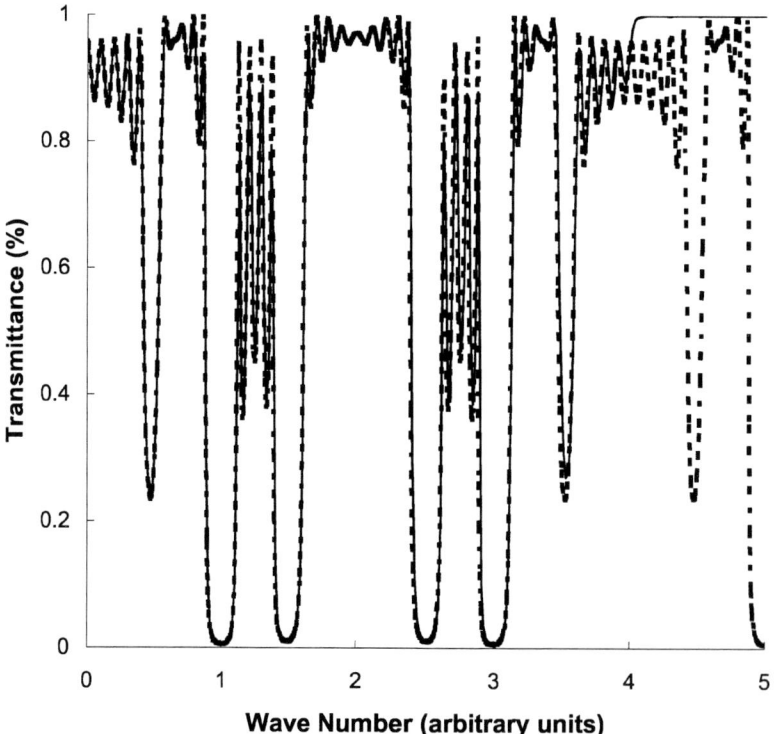

Figure 2.24(b) The transmittance of rugate TMD (solid) and discrete-layered TMD (dashed) in Sec. 2.8.3.

Chapter 3

Discrete Apodization
of TMDs

Chapter 2 discussed the basic form of TMDs for discrete-layer, homogeneous designs and inhomogeneous rugate TMDs. For rugate designs, envelope functions are used to define the refractive-index profile of the rugate design. Typically, rugate designs with these envelope functions have reduced or suppressed reflected-ripple structure in their passbands, hence apodization.[16] Discrete-layered TMDs can be apodized in a similar fashion except that the ripple structure of TMD passbands is not suppressed like apodized rugate designs. However, the apodization of TMDs does change the spectral performance of TMDs. The purpose of this chapter is to briefly investigate the relative stopband spectral positions of TMDs with the application of various envelope functions.

3.1 Introduction

Equation (1.23) is an example of an apodizing envelope function that is applied to a rugate coating design. A method of TMD apodization was reported[24] where an amplitude modulation (AM) based scheme was applied to a TMD. Here, the modulation amplitude of the TMD, k, was replaced by the function

$$k = k'm(L) \tag{3.1}$$

and

$$m(L) = \sin(2\pi f_1 L), \tag{3.2}$$

where $m(L)$ is the envelope function that is defined by its own modulation amplitude and frequency, k_1 and f_1, respectively. The insertion of Eqs. (3.1) and (3.2) into Eq. (2.15) gives

$$T(L) = T_{AVG}[1 + k' \sin(2\pi f_1 L) \cos(2\pi f L)]. \tag{3.3}$$

For Eq. (3.2) to perform as a typical envelope function, f_1 must be much smaller than the modulation frequency f shown in Eq. (3.3). The next section covers examples of this, and when

$$f_1 \to f. \tag{3.4}$$

3.2 Amplitude Modulation

Equation (3.3) determines the layer thickness for an amplitude-modulated TMD (or AM-TMD). In this chapter, several examples of AM-TMDs are discussed where the AM frequency f_1 is much smaller than the TMD modulation frequency f, and where f_1 is based on the number of layers selected *a priori* for the AM-TMD. First, a simple envelope is generated when the AM frequency f_1 is defined to be

$$f_1 = \frac{1}{2L_{TOT}}. \tag{3.5}$$

In Eq. (3.5), f_1 is determined by the total number of layers of the AM-TMD, or L_{TOT}. Here, the TMD's modulation amplitude defines the layer-thickness profile (envelope) for the example AM-TMD shown in Fig. 3.1. Figure 3.2 shows the corresponding spectral performance.

Next, Fig. 3.3 shows all possible stopbands for several TMDs with different modulation frequencies. In this graph, the same parameter set was used as that for Figs. 3.1 and 3.2. The TMDs for all stopbands shown in Fig. 3.3 have an AM frequency of 0.01 and 50 total layers (for each design). Again, the spectral center of each stopband is defined by Eq. (2.23). Additionally, the bandwidth of most stopbands shown in Fig. 3.3 is generally larger than the ones shown in Fig. 2.9 for TMDs without AM.

3.2.1 AM frequencies that approach TMD modulation frequencies

The above discussion was limited to the case defined by Eq. (3.5). When the AM frequency is increased and it approaches the TMD modulation frequency, all possible stopbands are found for three cases of AM frequency. Here, the three arbitrarily selected AM frequencies are 0.02, 0.04, and 0.125.

First, the layer-thickness profiles are presented in Figs. 3.4(a) to (c) for the three AM-TMDs mentioned above. Next, all possible stopbands are shown in Figs. 3.5 to 3.7 as a function of modulation frequency. Here, the spectral positions of all stopbands are generally determined by Eq. (2.23). However, the pattern of stopbands in Fig. 3.7 departs from what Eq. (2.23) predicts. Figures 3.8 and 3.9 are the same as Figs. 3.5 to 3.7 except the AM frequencies are 0.1 and 0.3, respectively, and the spectral range is plotted to three wave numbers instead of five. As observed from Figs. 3.7 to 3.9, the resulting (general) pattern of stopbands is also a function of the AM frequency. From these graphs, the USE for TMDs, Eq. (2.23), can be rewritten for AM-TMDs as Eq. (3.6). However, Eq. (3.6) is approximate because more complex patterns of stopbands are present, especially in Figs. 3.8 and 3.9:

$$\sigma_{M,N} \approx \sigma_0[2N(f - f_1) + (2M - 1)]. \tag{3.6}$$

Equation (3.6) becomes less accurate when the AM frequency f_1 approaches the TMD frequency f.

When the AM and TMD frequencies are equal, Eq. (3.6) does not approximate the center of the stopband positions. In this situation, the AM-TMD is equavalent to a TMD without AM. In this case, Eq. (2.23) determines the spectral center of all possible stopbands.

3.3 Gaussian Envelope Functions

As an alternative to an AM of the TMD envelope function, this section investigates the use of Gaussian envelope functions. Equation (3.2) is replaced with the Gaussian function

$$m(L) = e^{-B(L\frac{L_{TOT}}{2})^2}. \tag{3.7}$$

Here, L is the layer number, L_{TOT} is the total number of layers in the TMD, and B is arbitrarily selected. For example, three TMDs with Gaussian envelope functions are shown in Fig. 3.10 for different values of B for a TMD with 50 layers.

The layer-thickness profile and spectral reflectance of an arbitrary TMD with a Gaussian envelope profile is shown in Figs. 3.11 and 3.12, respectively. Here, the passband ripple structure is similar to that of the AM-TMDs shown in Fig. 3.1.

Figures 3.13 to 3.16 correspond to the same Gaussian-enveloped TMD as above except B is increased from 0.01 to 0.05 and 0.25. Figures 3.13 and 3.15 show that increasing the value of B reduces the overall modulation of the layer thickness. Also, ripple structure in the passbands increases with larger values of B. For values of B that are approximately 2 or larger, the film thickness profile of the TMD degenerates to essentially a Fabry-Perot cavity design with one half-wave (HWOT) layer at or near the center of the stack, and the remaining layers with thickness of one quarter-wave (QWOT).

Thickness-modulated designs with Gaussian envelope functions (G-TMD) provide some useful spectral performance variations from AM-TMDs. Figures 3.17 to 3.19 show the spectral positions of stopbands for several G-TMDs as a function of modulation frequency. Unlike the stopbands shown in Figs. 3.3 and 3.5 for AM-TMDs, these stopbands generally have increased bandwidth, especially for the (1,0)- and (2,0)-order stopbands at 1 and 3 μm^{-1}. Also note that the bandwidth of the nonharmonic stopbands is generally decreased.

3.4 Application

3.4.1 Multiband reflective and transmissive filter

The following fictitious specifications are proposed for a thin-film design:

Passband 1:   Transmittance $>95\%$, $\lambda = 532\,nm$
Stopband 1:   Reflectance $>98\%$, $\lambda = 605–635\,nm$
Stopband 2:   Reflectance $>99.5\%$, $\lambda = 925–1203\,nm$

First, the TMD design method in Chapter 2 is evaluated to see if a TMD will achieve these specifications. The USE-TMD, Eq. (2.22) or (2.23), is employed to determine the spectral spacing of two stopbands. The center wavelengths of the above stopbands, 1 and 2, are 620 and 1064 nm, respectively. The (1,0)- and (1,1)-order TMD stopbands are selected, so the TMD modulation frequency is calculated to be 0.358. The bandwidth ratios of stopbands 1 and 2 are 1.05 and 1.3, respectively. A TMD has the maximum width of the (1,0) stopband when the modulation amplitude is low. Therefore, Fig. 2.8 is evaluated (modulation amplitude = 0.25) for the stopband ratios of the (1,0) and (1,1) stopbands for the modulation frequency 0.358. In this case, the stopband ratio for (1,1) is 1.09, which meets the specification requirement for stopband 1. However, the stopband ratio for (1,0) is only 1.24, which does not meet the specification for stopband 2. There is no stopband present at the required passband.

Next, AM-TMDs are evaluated as a possible design to achieve the above specifications. For AM-TMDs, lower AM frequencies have, in general, larger bandwidth stopbands, as observed in Figs. 3.5 to 3.7. Using a modulation amplitude frequency of 0.02 (and the same parameters as those for Fig. 3.5), the bandwidth ratios are calculated for the AM-TMD. For the same TMD modulation frequency of 0.358, the stopband ratios are 1.16 and 1.32, respectively, for stopbands 1 and 2. Therefore, this design meets both requirements, and no stopband is present at the passband.

A trial AM-TMD design is then determined from Eq. (3.3). Figure 3.20 shows the layer-thickness profile of this AM-TMD, and Fig. 3.21 shows the AM-TMD's reflectance versus wavelength. Figure 3.21 shows that the reflectance specifications are virtually met, or could be met with additional layers, and the passband exists at 532 nm. With the addition of layers to this design and by refining via optimization software to reduce ripples at 532 nm, this design would meet the stated specifications.

3.5 Exercises

1. Determine the general analytical relationship between the bandwidth of stopbands with and without AM as defined by Eq. (3.5).
2. Write a software program (or modify an existing program) to determine all stopbands for AM-TMDs.
3. Using a polynomial or other function, determine the relationship between the bandwidth of stopbands produced by G-TMDs as a function of B from Eq. (3.7) and the TMD's modulation frequency.
4. For the example application in Sec. 3.4.1, explain why the spectral center of the AM-TMD for the (1,0)-order TMD stopband is off-center from the target wavelength (i.e., 1064 nm).

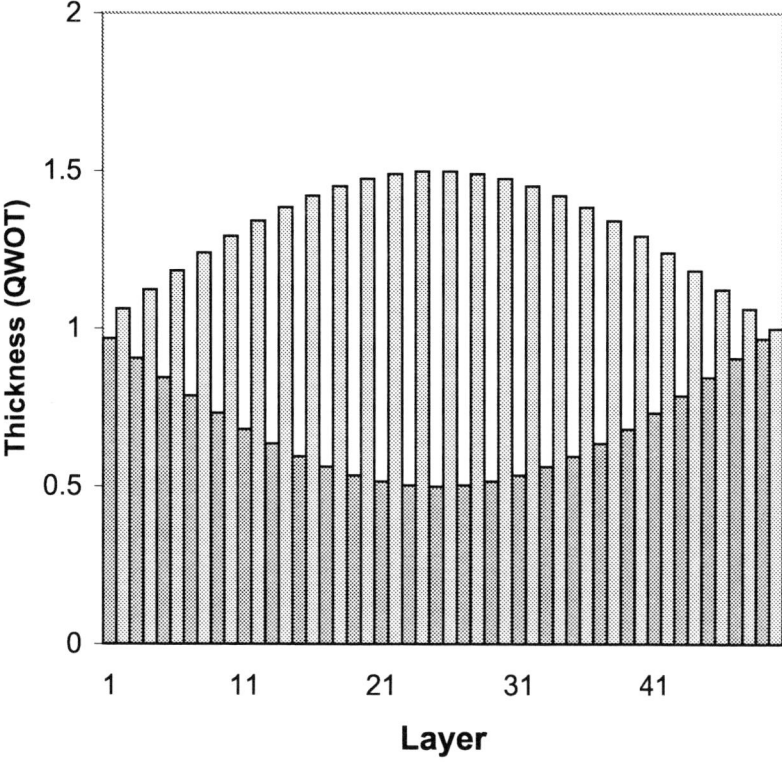

Figure 3.1 Layer-thickness profile for an AM-TMD with parameters $f = 0.5$, $f_1 = 0.01$, $k = 0.5$, and $L = 50$.

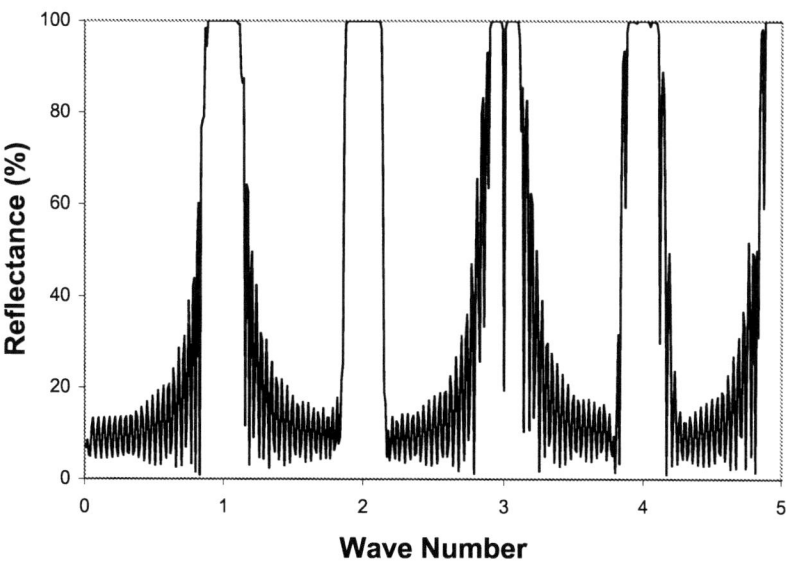

Figure 3.2 Spectral reflectance of the AM-TMD defined in Fig. 3.1 with refractive indices of 1.46 and 2.25 for the films and 1.52 for the substrate.

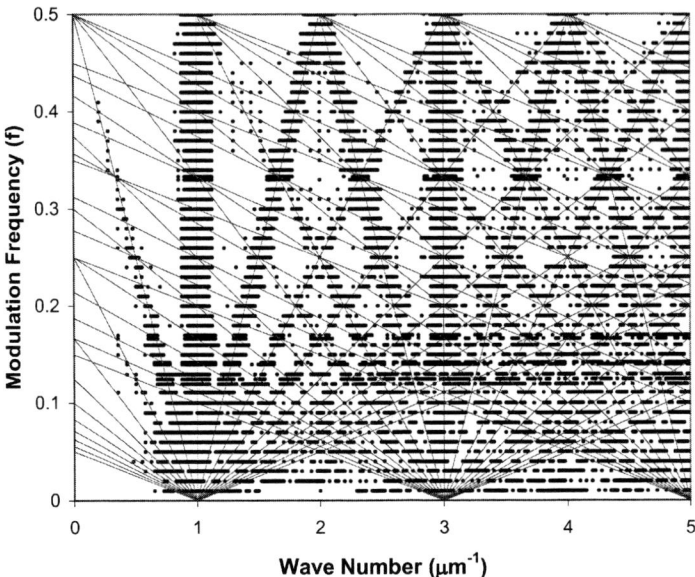

Figure 3.3 Spectral positions of all stopbands within the wave number range of 0–5 for an AM-TMD with the following parameters: refractive index of films, 1.46, 2.25; substrate, 1.52; modulation amplitude, 0.5; AM frequency, 0.01; and average layer thickness, 1 μm QWOT. Compare to Fig. 2.9 with a constant modulation amplitude of 0.5.

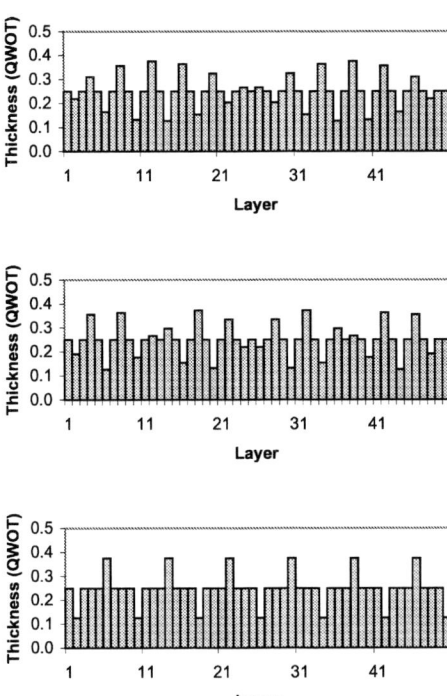

Figures 3.4(a)–(c) Layer-thickness profiles for three AM-TMDs with AM frequencies of 0.02, 0.04, and 0.125.

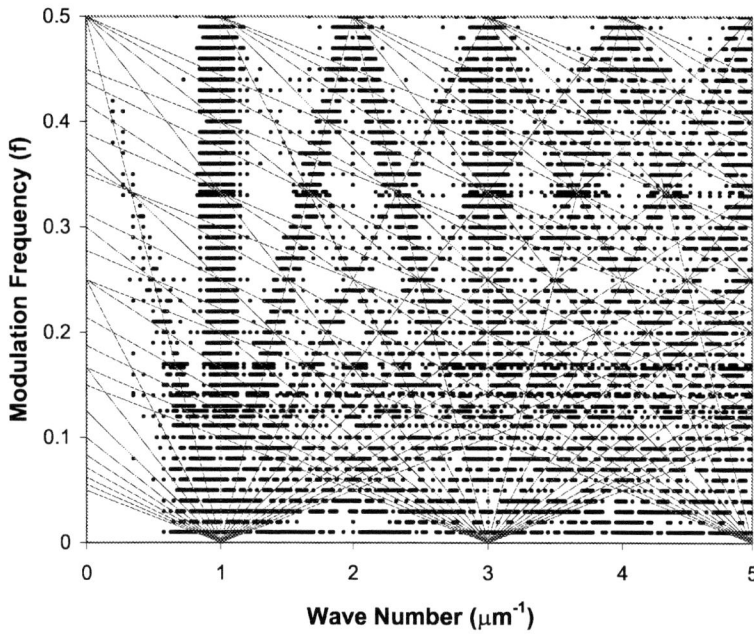

Figure 3.5 The spectral positions of all stopbands within the wave number range of 0–5 for an AM-TMD with the following parameters: refractive index of films, 1.46, 2.25; substrate, 1.52; modulation amplitude, 0.5; AM frequency, 0.02; and average layer thickness, 1 μm QWOT.

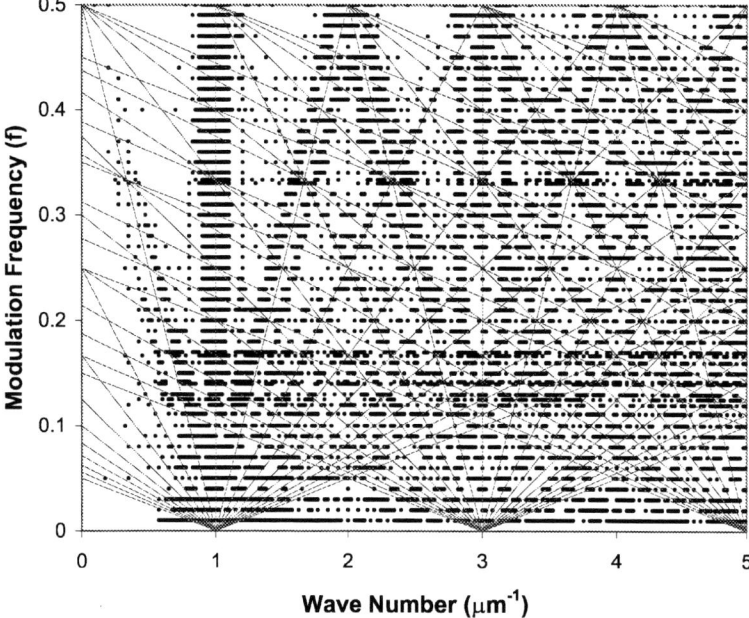

Figure 3.6 Spectral positions of all stopbands within the wave number range of 0–5 for an AM-TMD with the following parameters: refractive index of films, 1.46, 2.25; substrate, 1.52; modulation amplitude, 0.5; AM frequency, 0.04; and average layer thickness, 1 μm QWOT.

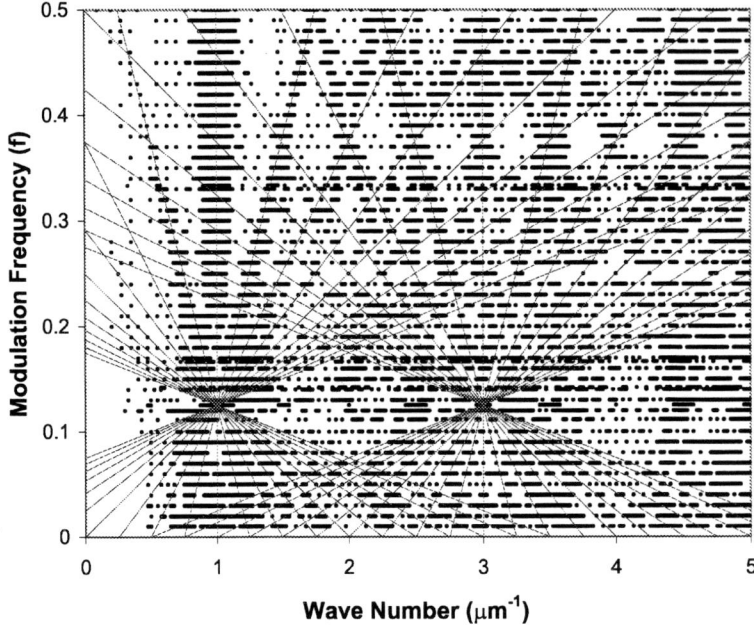

Figure 3.7 Spectral positions of all stopbands within the wave number range of 0–5 for an AM-TMD with the following parameters: refractive index of films, 1.46, 2.25; substrate, 1.52; modulation amplitude, 0.5; AM frequency, 0.125; and average layer thickness, 1 μm QWOT.

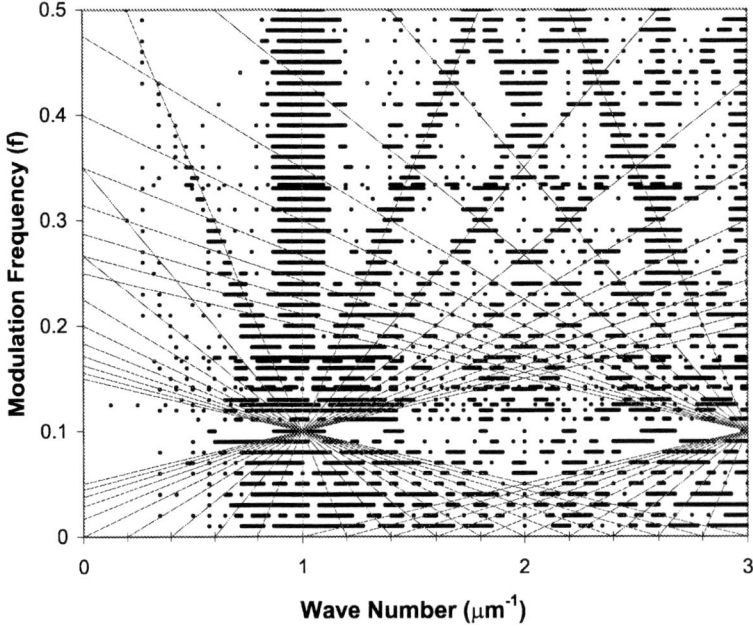

Figure 3.8 Spectral positions of all stopbands within the wave number range of 0–5 for an AM-TMD with the following parameters: refractive index of films, 1.46, 2.25; substrate, 1.52; modulation amplitude, 0.5; AM frequency, 0.1; and average layer thickness, 1 μm QWOT.

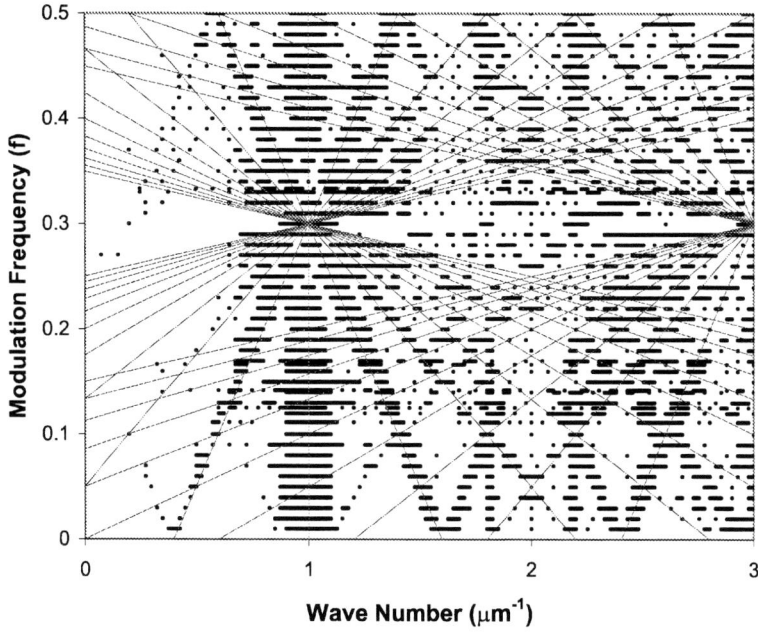

Figure 3.9 Spectral positions of all stopbands within the wave number range of 0–5 for an AM-TMD with the following parameters: refractive index of films, 1.46, 2.25; substrate, 1.52; modulation amplitude, 0.5; AM frequency, 0.3; and average layer thickness, 1 μm QWOT.

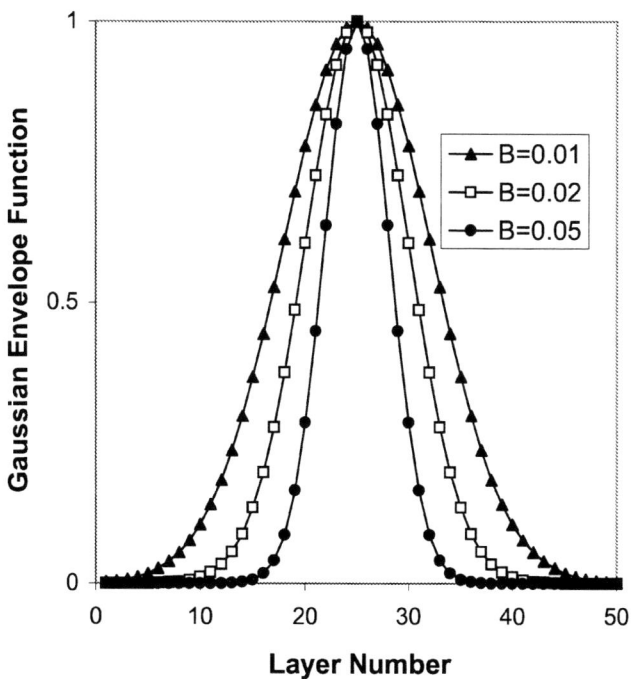

Figure 3.10 Three examples of Gaussian envelope functions for a 50-layer TMD.

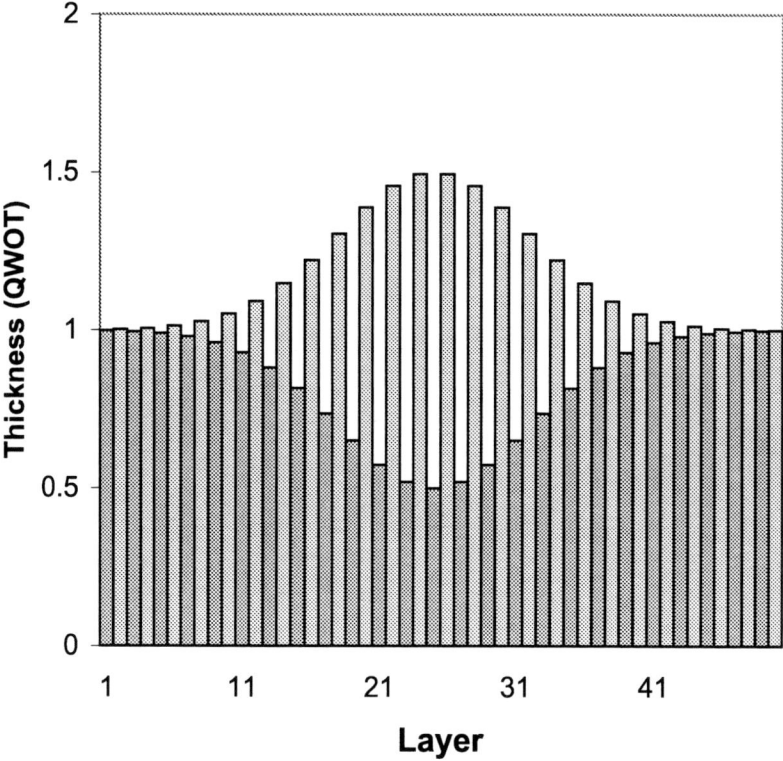

Figure 3.11 Layer-thickness profile for a Gaussian-envelope TMD with parameters $f = 0.5$, $f_1 = 0.01$, $k = 0.5$, $B = 0.01$, and $L = 50$.

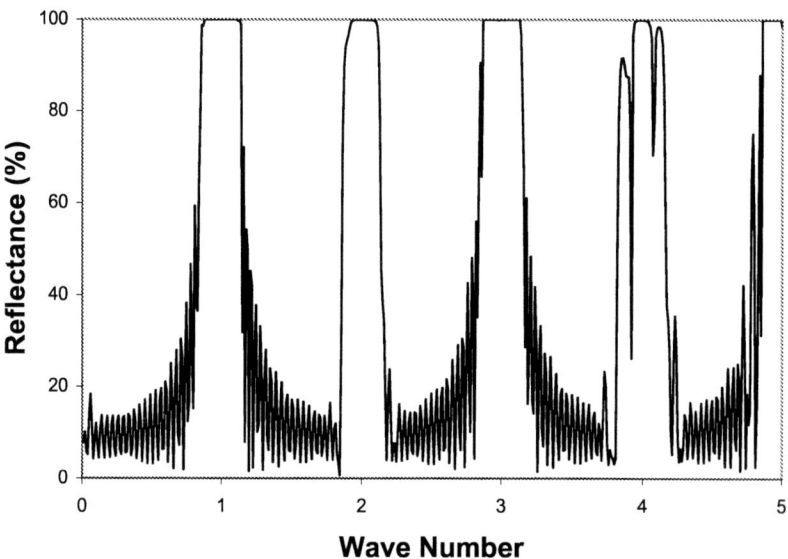

Figure 3.12 Spectral reflectance of a TMD with a Gaussian envelope function de-fined in Fig. 3.11, and having refractive indices of 1.46 and 2.25 for the films and 1.52 for the substrate.

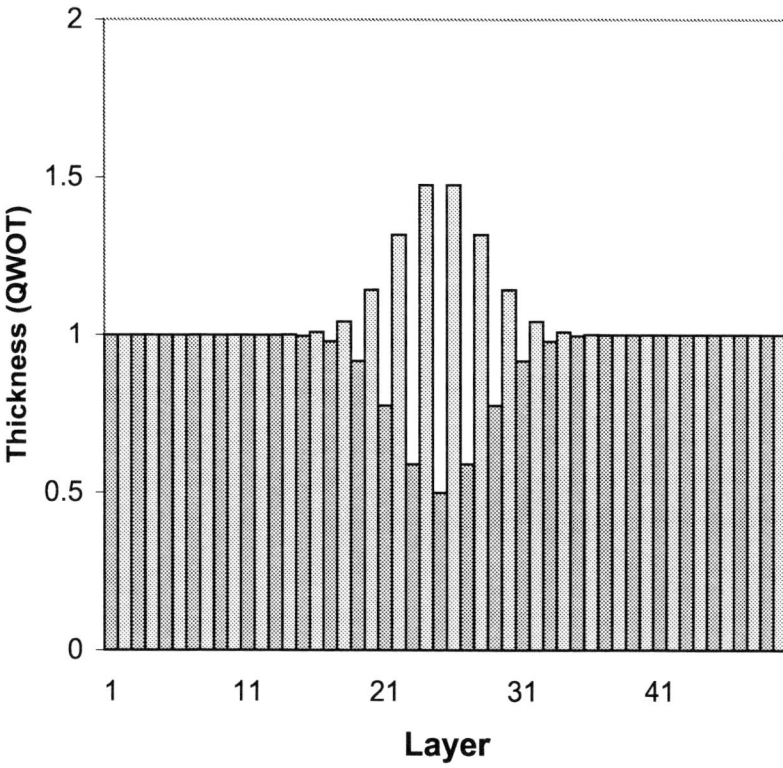

Figure 3.13 Layer-thickness profile for a Gaussian-envelope TMD with parameters $f = 0.5$, $f_1 = 0.01$, $k = 0.5$, $B = 0.05$, and $L = 50$.

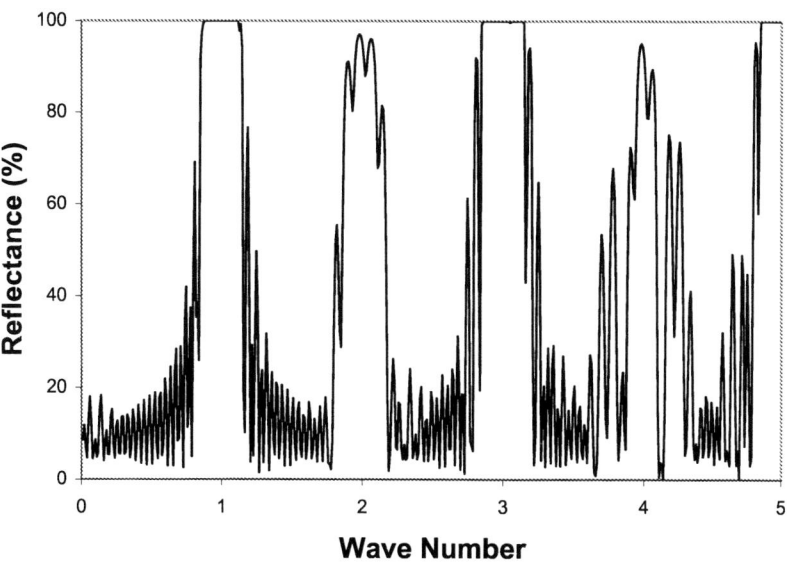

Figure 3.14 Spectral reflectance of a TMD with a Gaussian envelope function defined in Fig. 3.13, and having refractive indices of 1.46 and 2.25 for the films and 1.52 for the substrate.

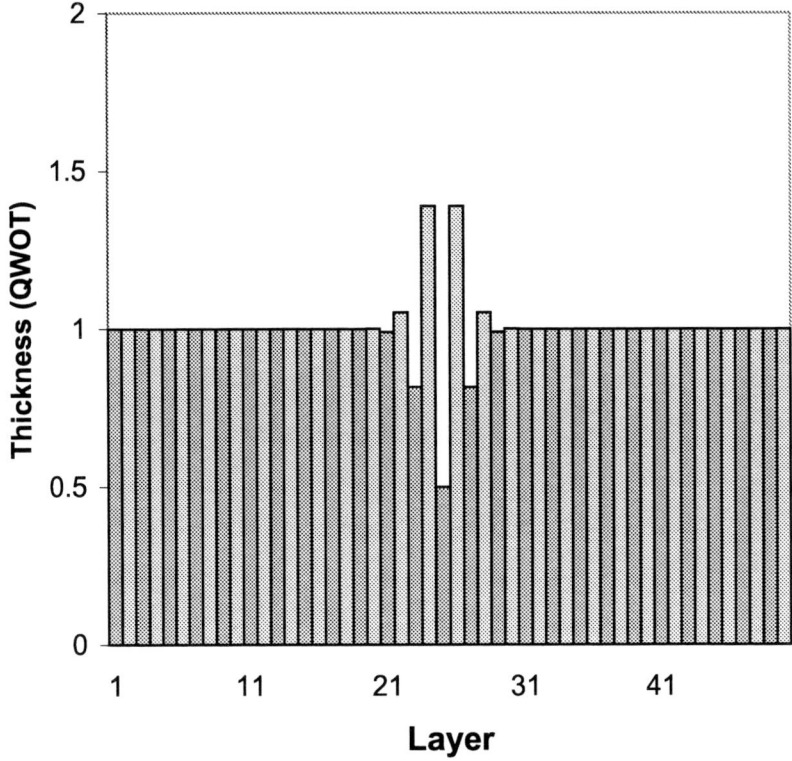

Figure 3.15 Layer-thickness profile for a Gaussian-envelope TMD with parameters $f = 0.5$, $f_1 = 0.01$, $k = 0.5$, $B = 0.25$, and $L = 50$.

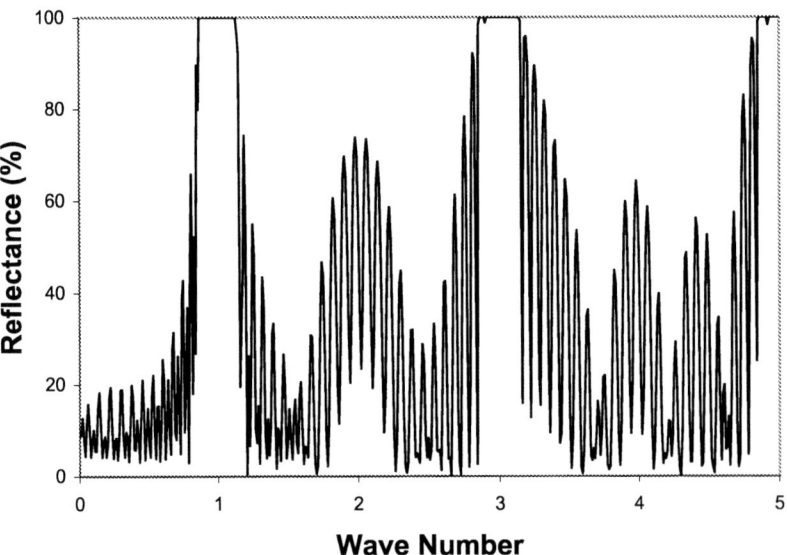

Figure 3.16 Spectral reflectance of a TMD with a Gaussian envelope function defined in Fig. 3.15 and having refractive indices of 1.46 and 2.25 for the films and 1.52 for the substrate.

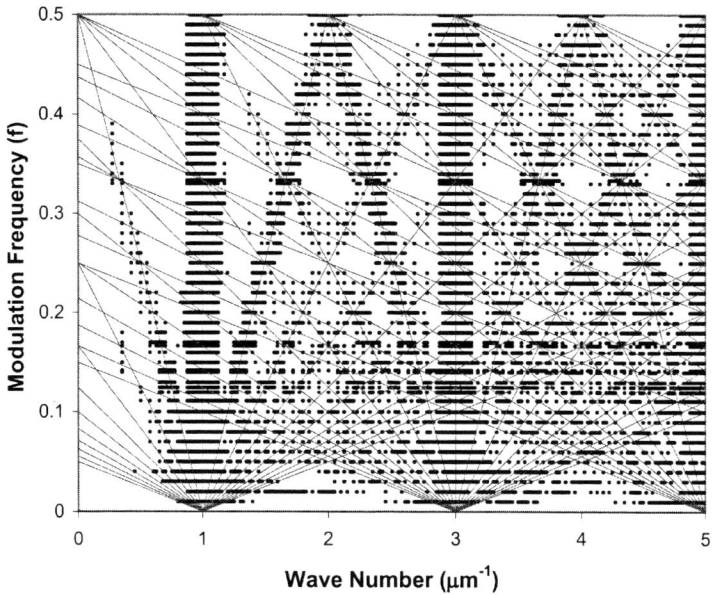

Figure 3.17 Spectral postions of all stopbands within the wave number range of 0–5 for a TMD with a Gaussian envelope function and having the following parameters: refractive index of films, 1.46, 2.25; substrate, 1.52; modulation amplitude, 0.5; Gaussian parameter, $B = 0.01$; and average layer thickness, 1 μm QWOT. Compare to Fig. 3.3 with a constant modulation amplitude of 0.5.

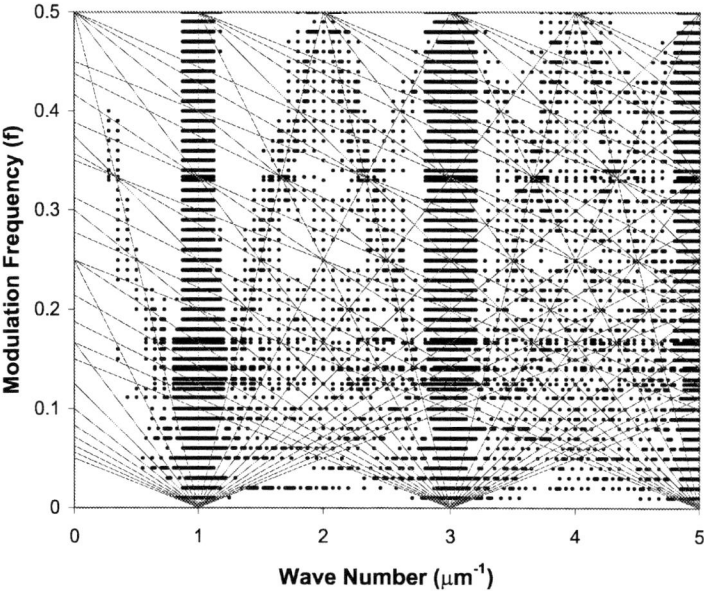

Figure 3.18 Spectral postions of all stopbands within the wave number range of 0–5 for a TMD with a Gaussian envelope function and having the following parameters: refractive index of films, 1.46, 2.25; substrate, 1.52; modulation amplitude, 0.5; Gaussian parameter, $B = 0.05$; and average layer thickness, 1 μm QWOT. Compare to Fig. 3.3 with a constant modulation amplitude of 0.5.

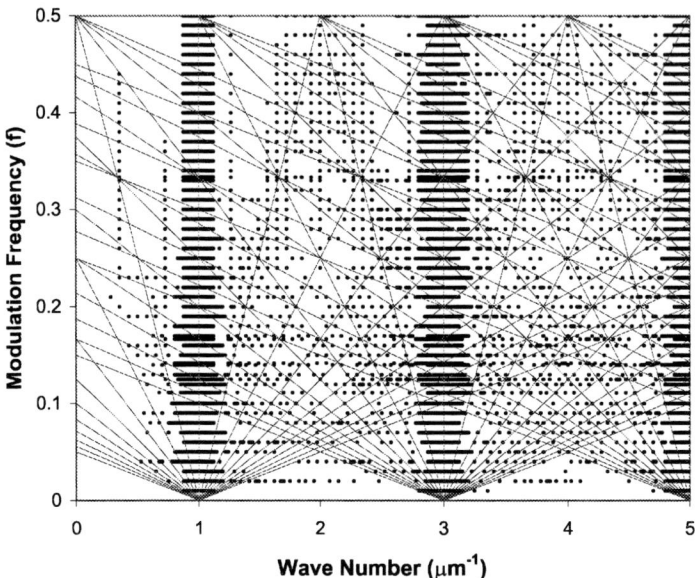

Figure 3.19 Spectral postions of all stopbands within the wave number range of 0–5 for a TMD with a Gaussian envelope function and having the following parameters: refractive index of films, 1.46, 2.25; substrate, 1.52; modulation amplitude, 0.5; Gaussian parameter, $B = 0.25$; and average layer thickness, 1 μm QWOT. Compare to Fig. 3.3 with a constant modulation amplitude of 0.5.

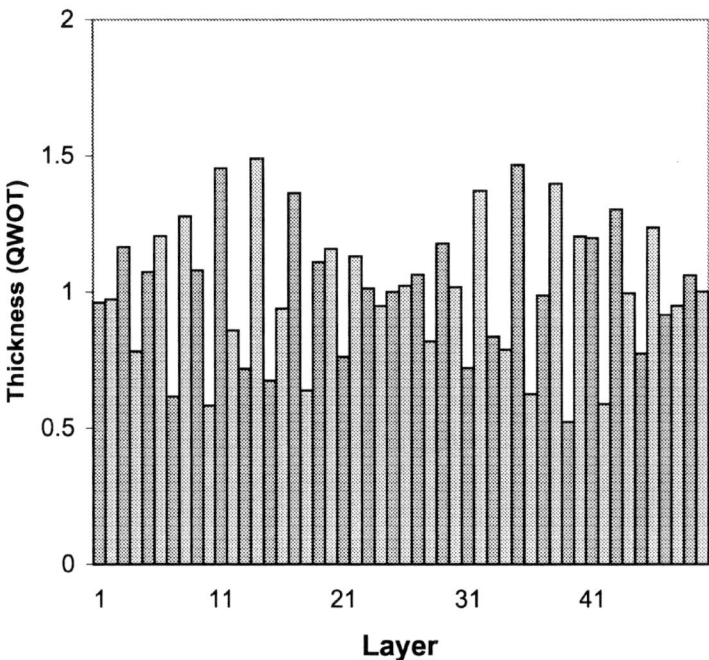

Figure 3.20 Layer-thickness profile for an AM-TMD with parameters $f = 0.358$, $f_1 = 0.02$, $k = 0.5$, and $L = 50$. All refractive indices are the same as these used in Figs. 3.1 and 3.2.

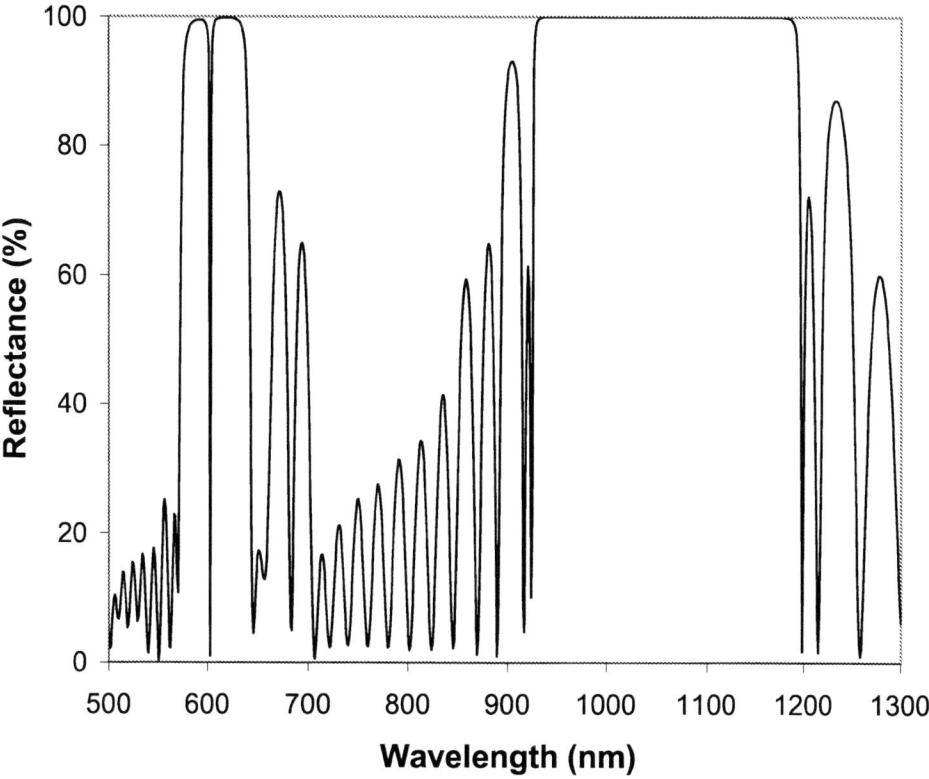

Figure 3.21 Spectral reflectance of the AM-TMD described in Fig. 3.20. Note that stopbands are present in the ranges of 605–635 nm and 925–1200 nm, and ripples exist in the passband near 532 nm.

Chapter 4

Other Complex TMDs

Lasers that produce femtosecond pulses require dispersion-controlled mirrors. Several researchers have reported on thin-film design methods for these mirrors that are similar to TMDs that produce a stopband. These mirror designs have increased bandwidth over quarter-wave stacks and phase control of the reflected group delay, or GD [also group-delay dispersion (GDD) or wavelength-dependent GDD[34]]. For example, Szipöcs et al. reported the use of a monotonically varying quarter-wave stack (starting design) that is optimized to produce a required GDD as a function of wavelength.[35] This monotonically varying stack is referred to as a chirped coating and is similar to progressive series stacks (see Sec. 2.3.2). Tempea et al. reported the use of a layer-thickness modulation method for chirped mirror coatings.[36] This method generates a starting design with improved GDD by chirping the period of the modulated layers while simultaneously increasing the modulation amplitude. In this case, less optimization is required to adjust or smooth the GD of the starting design to target values. Matuschek et al. reported on double-chirped mirrors where more sophisticated analytical design methods produce starting designs that are very close to the desired reflectance and GD.[37] Here, double-chirped starting designs have several features: independently chirped, low and high refractive-index layers; specific groups of layers chirped for impedance matching; and an impedance-matched, broadband antireflection (AR) coating placed on top of the mirror coating to further suppress oscillations in the GD. This starting design requires minimal optimization to improve its performance.

Southwell reported on the design of rugate filters using wavelet design methods to increase the bandwidth of the stopband.[38] A wavelet is defined as a fully apodized sine wave refractive-index profile (see Fig. 1.6). This method increases the bandwidth of the resulting stopband by using several overlapping wavelets. These wavelet designs are free of stopband harmonics and reflection ripples in the passbands. Note that consideration of GD is not part of this design method.

In all of the above design methods, the objective is to accomplish one or both of the following: (1) to widen the bandwidth and increase the reflectance of the first-order stopband; and (2) to achieve the desired GD (and GDD) versus wavelength. This chapter starts by examining chirped TMDs from a more general perspective, like Chapters 2 and 3, but for only the spectral properties of multiple stopbands. Next, the GD of a chirped-TMD design is briefly discussed. This chapter concludes by considering a half-modulation TMD function and its effect on stopband positions and GD performance. The limitations of both methods are also addressed.

4.1 Chirped TMD Modulation

First, the spectral performance of chirped TMDs is evaluated for multiple stopbands and for the bandwidth of the produced stopbands. For chirped modulation, the modulation frequency is swept or varied over a finite range. One possible sweep function is a linear sweep given by

$$T(L) = T_{AVG}[1 + k\cos(2\pi fL)], \tag{4.1}$$

where f is also a function of L and is given by

$$f(L) = f_b + k_c(L - 1). \tag{4.2}$$

Linear sweep - function

In Eq. (4.2), f_b is the TMD modulation frequency at the first layer, and k_c is the chirp or sweep constant. For a given chirped TMD with L layers and a desired modulation frequency range of f_b to $f_{b'}$, the chirp constant k_c can be determined by Eq. (4.3):

$$k_c = \frac{f_{b'} - f_b}{L_{TOT} - 1}. \tag{4.3}$$

Again, as with previously discussed TMDs, the modulation amplitude k in Eq. (4.1) varies between 0 and 1, inclusive, and L_{TOT} is the total number of layers in the TMD. Figure 4.1 shows the layer-thickness profile of an example sweep or chirp function TMD (C-TMD) that was determined from Eqs. (4.1) to (4.3).

4.1.1 Degenerate cases of C-TMD

Before investigating the performance of C-TMDs, some degenerate cases are evaluated for further insight to TMDs. Figures 4.2 to 4.4 show the stopband positions as a function of the modulation amplitude k and spectral frequency. For all three cases, the modulation frequency is held constant. Hence, these designs are TMDs as discussed in Chapter 2.

The first group of designs, where $f = 0.5$, is also a degerate TMD, where stopbands are present at integer multiples of the first-order stopband only. Starting at $k = 0.0$, the classic stopbands are present in Fig. 4.2 for a 1:1 quarter-wave stack (equal layer thicknesses). All possible detuned quarter-wave stacks are generated as k varies from 0 to 1.

The next two design groups demonstrate the nonharmonic stopbands of TMDs. Figures 4.3 and 4.4 show several groups of stopbands that exist only at some values of k.

4.1.2 Selected cases of C-TMD

This section presents three cases of C-TMD that demonstrate typical performance. Figures 4.5 to 4.7 show the stopband positions for the three C-TMDs with modulation frequencies shown in Table 4.1.

Figure 4.5 shows how the positions (spectral frequency) change as a function of the modulation amplitude. Note that the C-TMD demonstrated in this figure produces relatively wide stopbands.

Table 4.1 C-TMD modulation frequencies and periods for selected designs.

Figure	$f_b(T)$	$f_{b'}(T)$
4.5	0.05 (20)	0.125 (8)
4.6	0.117647 (8.5)	0.125 (8.0)
4.7	0.001 (1000)	0.1 (10)

When there is a small difference in the modulation frequency range, f_b to $f_{b'}$, the C-TMD produces stopbands that are not as dependent on modulation amplitude. As shown in Fig. 4.6, the stopbands have very little dependence on modulation amplitude above $k \approx 0.25$.

For large differences in the modulation frequency range, the C-TMD produces stopbands that are strongly dependent on modulation amplitude. This is shown in Fig. 4.7, where the bandwidth of the stopbands is also wider than for the previous two cases of C-TMDs.

4.1.3 Example C-TMD (bandwidth of stopband and GD performance)

Two groups of stopbands in Fig. 4.7 appear to converge as the modulation amplitude decreases to about 0.25. The spectral performance of a C-TMD with modulation amplitude $k = 0.25$ is shown in Fig. 4.8. From Fig. 4.7, the stopband at wave number 1 has a bandwidth ratio of 1.375/0.75, or 1.833. The reflectance GD for this C-TMD is shown in Fig. 4.9. Note that several GD spikes are present without applying any of the design methods described above (e.g., simultaneously increasing the modulation amplitude or adding an impedance-matching AR coating). Although further investigation of GD for C-TMDs would be insightful, it is beyond the scope of this text.

4.1.4 Limitations and applications of C-TMD to design problems

Several additional examples of C-TMDs were investigated in order to determine the patterns (if any) that existed among the produced stopbands. The typical patterns of C-TMD stopbands were complex and not readily useable for analytical design of multiple stopbands. However, the C-TMD design can extend TMDs for single or possibly two stopbands with increased bandwidth.

4.2 Half-modulation TMDs

This final investigation of TMDs examines the effect of modulating every other layer of a quarter-wave stack while holding the skipped layers at the QWOT of the quarter-wave stack. The objective here is to evaluate the position of stopbands as a function of modulation amplitude k and spectral frequency.

The H-TMD equations for layer thickness are given by Eqs. (4.4) and (4.5) for the case when the even-numbered layers of the TMD are modulated [note that

Eq. (4.4) is the same as Eq. (2.15)]:

For even-numbered layers: $\quad T(L) = T_{AVG}[1 + k\cos(2\pi fL)] \quad$ (4.4)

and

For odd-numbered layers: $\quad T(L) = T_{AVG}.$ (4.5)

For this modulation scheme, either the high or low refractive-index layers could be modulated. Also, Eqs. (4.4) and (4.5) could be swapped for modulation of the odd-numbered layers. The following discussions refer to modulation of even-numbered layers.

Figures 4.10 to 4.12 show the stopbands as a function of modulation frequency and spectral frequency for the modulation amplitudes of 0.25, 0.5, and 0.75, respectively. Notice that the first-order stopband at wave number 1 is, in general, preserved for all modulation frequencies. When Fig. 4.10 is compared to Fig. 2.8, there are three observable differences:

1. Stopbands are present near the first-, third-, and fifth-order harmonic stopbands of the original quarter-wave stack. Also, virtually no stopbands are present near the second- and fourth-order harmonic frequencies (i.e, 2 and 3 μm^{-1}).
2. The pattern of stopbands is symmetric about the modulation frequency of $f = 0.25$. This occurs because of the even-layer modulation. In contrast, the stopbands in Fig. 2.8 are *asymmetric* about $f = 0.25$.
3. Several of the stopband sequences (as a function of f) originate from even (second, fourth) harmonic frequencies when $f = 0.0$. This is in addition to the stopband sequences originating at $f = 0.0$ for the odd harmonic frequencies for the TMDs reported in Chapter 2. Hence, the patterns originating at $f = 0.0$ for even harmonic frequencies are not predicted by Eq. (2.23).

Figures 4.11 and 4.12 show similar patterns for TMDs, where an increased modulation amplitude increases the bandwidth of the nonharmonic stopbands and decreases the bandwidth of the harmonic stopbands. Again, these H-TMDs are symmetric about the modulation frequency of $f = 0.25$ for the modulation of even-numbered layers.

4.2.1 Example H-TMD (bandwidth of stopband and GD performance)

Several groups of stopbands in Fig. 4.11 converge as the modulation frequency approaches 0 at wave number 1. The reflectance of an H-TMD with the parameters $f = 0.01$, $k = 0.5$, and $L_{TOT} = 50$ is shown in Fig. 4.13. Note that the spectral centering of the stopband was changed from 1000 nm (i.e., 1 μm^{-1} wave number) to $T_{AVG} = 800$ nm. This design has its even-numbered layers modulated; these layers were also selected to be the high refractive index. The reflectance shown in Fig. 4.13 is near unity over the wavelength range of approximately 600 to 1100 nm.

This wavelength range is typical for dispersion-controlled mirror coatings designed for Ti:sapphire lasers.

Next, the reflectance GD for this H-TMD is shown in Fig. 4.14. Note that several GD spikes are present without the application of design methods used by other researchers or the optimization of the design (e.g., simultaneously increasing the modulation amplitude or adding an impedance-matching AR coating). However, this design generally has a negative GD, where the GD decreases with the wave number or frequency (desired). With some optimization, the GD of this H-TMD can be smoothed while preserving reflectance, as shown in Figs. 4.13 and 4.14. As other researchers have observed, very little adjustment or refinement of the layer thickness is needed on a good starting design. Figure 4.15 shows the layer thickness of the H-TMD (starting design) and the optimized design. Further performance improvements can be achieved by applying the design methods reported.[35-37]

4.2.2 Limitations and applications of H-TMD to design problems

The symmetry of stopband positions about the modulation frequency of 0.25 provides no unique solutions for $f > 0.25$. However, additional stopbands are present near the odd harmonics that could be used for several wavelengths in close proximity. Additionally, stopbands are not present near the even harmonics, so designs that require passbands in this region could use the H-TMD instead of the TMD.

4.3 Exercises

1. For C-TMDs with given refractive indices, what modulation frequencies and amplitude produce stopbands with the largest bandwidth?
2. Based on the empirical stopband position data in Figs. 4.8 to 4.10, determine an equation that predicts the stopband positions for H-TMDs (for modulation of even-numbered layers). For reference, see Eq. (2.23).
3. Write or modify a computer program that will calculate the spectral positions of stopbands for H-TMDs for modulation of both even- and odd-numbered layers.
4. Based on the results of Exercise 3, what is the resulting pattern of stopbands for modulating the odd-numbered layers? How does this differ from the even-layer modulation?
5. Modify the program from Exercise 3 to determine all possible stopbands for the rectified modulation scheme given by $T(L) = T_{AVG}[1 + k|\cos(2\pi f L)|]$. Here, every other layer is modulated. What is the resulting pattern of stopbands?

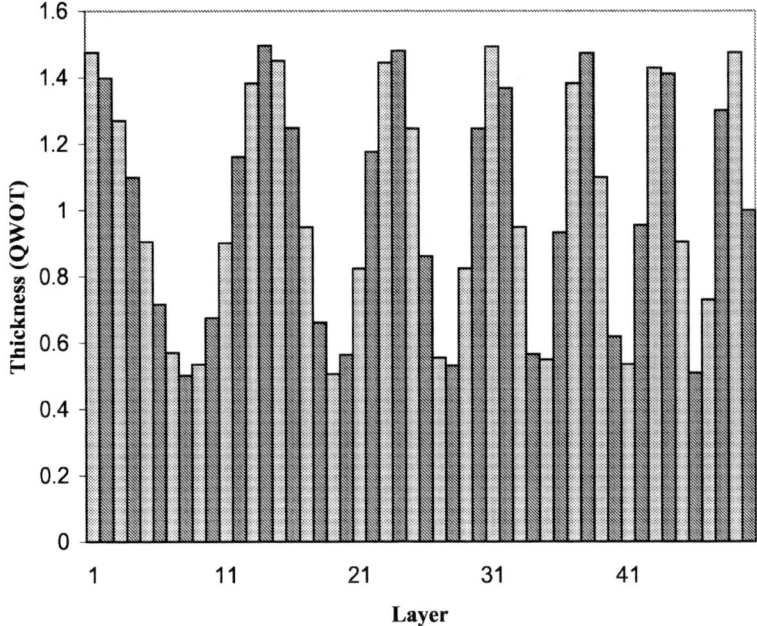

Figure 4.1 Layer-thickness profile for a chirped TMD with the following parameters: $f_b = 0.05$, $f_{b'} = 0.125$, $k = 0.5$, and $L = 50$.

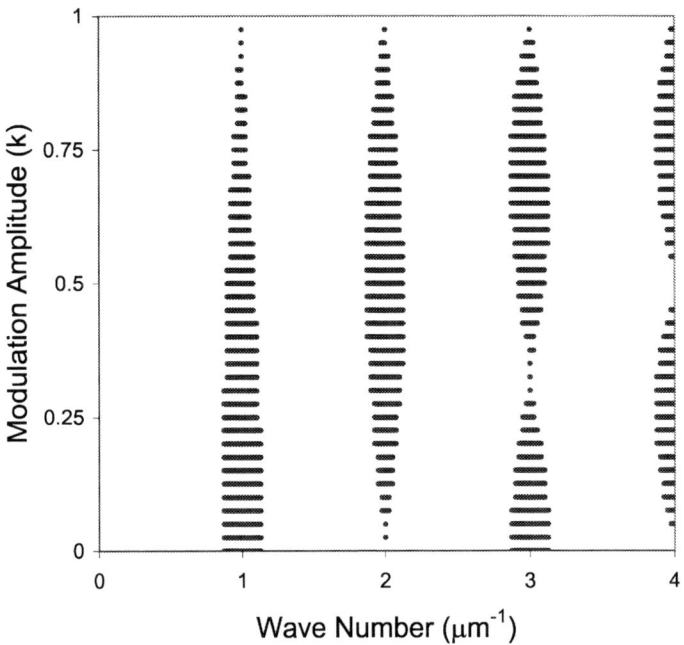

Figure 4.2 Spectral positions of all stopbands for a modulation frequency of 0.5. For all designs, the refractive indices for the ambient, substrate, and two films are 1.0, 1.52, 1.46, and 2.25, respectively.

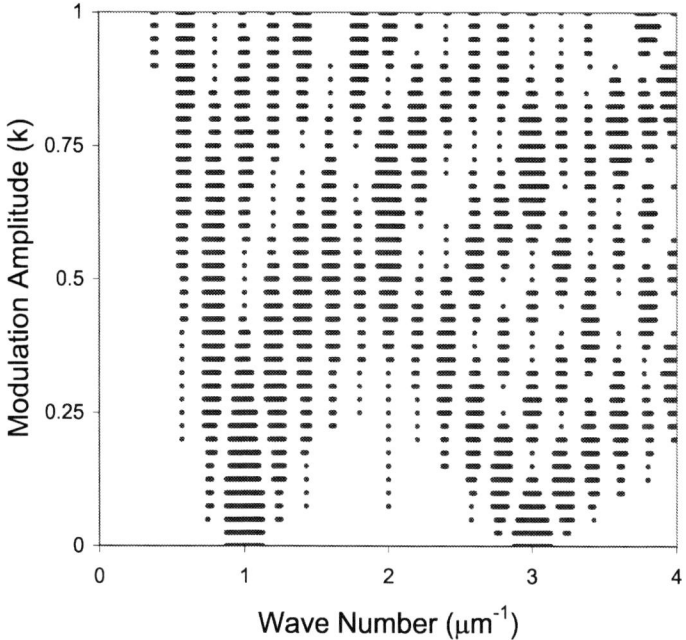

Figure 4.3 Spectral positions of all stopbands for a modulation frequency of 0.1. For all designs, the refractive indices for the ambient, substrate, and two films are 1.0, 1.52, 1.46, and 2.25, respectively.

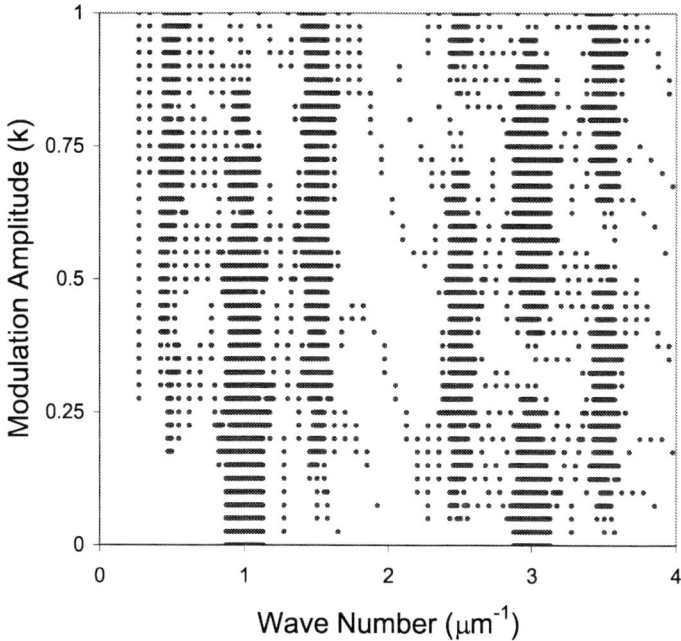

Figure 4.4 Spectral positions of all stopbands for a modulation frequency of 0.25. For all designs, the refractive indices for the ambient, substrate, and two films are 1.0, 1.52, 1.46, and 2.25, respectively.

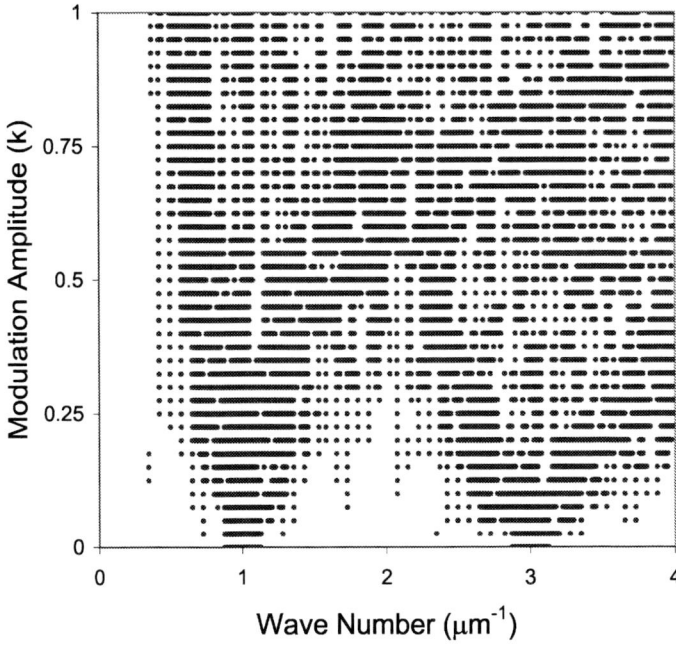

Figure 4.5 Spectral positions of all stopbands for $f_b = 0.05$ and $f_{b'} = 0.125$. For all designs, the refractive indices for the ambient, substrate, and two films are 1.0, 1.52, 1.46, and 2.25, respectively.

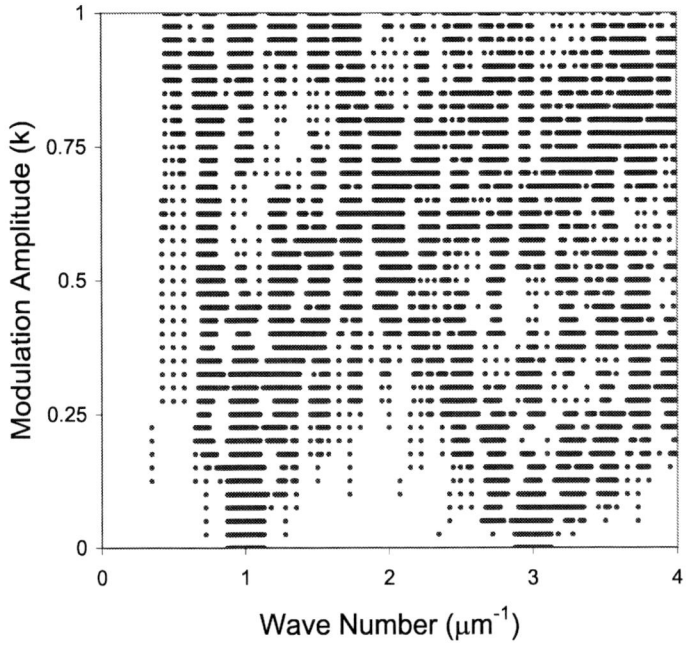

Figure 4.6 Spectral positions of all stopbands for $f_b = 1/8.5$ and $f_{b'} = 1/8$. For all designs, the refractive indices for the ambient, substrate, and two films are 1.0, 1.52, 1.46, and 2.25, respectively.

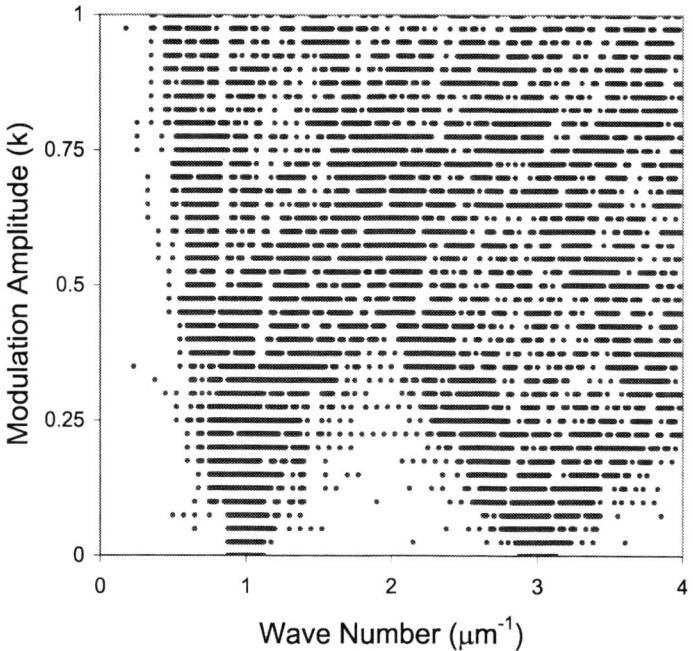

Figure 4.7 Spectral positions of all stopbands for $f_b = 0.001$ and $f_{b'} = 0.1$. For all designs, the refractive indices for the ambient, substrate, and two films are 1.0, 1.52, 1.46, and 2.25, respectively.

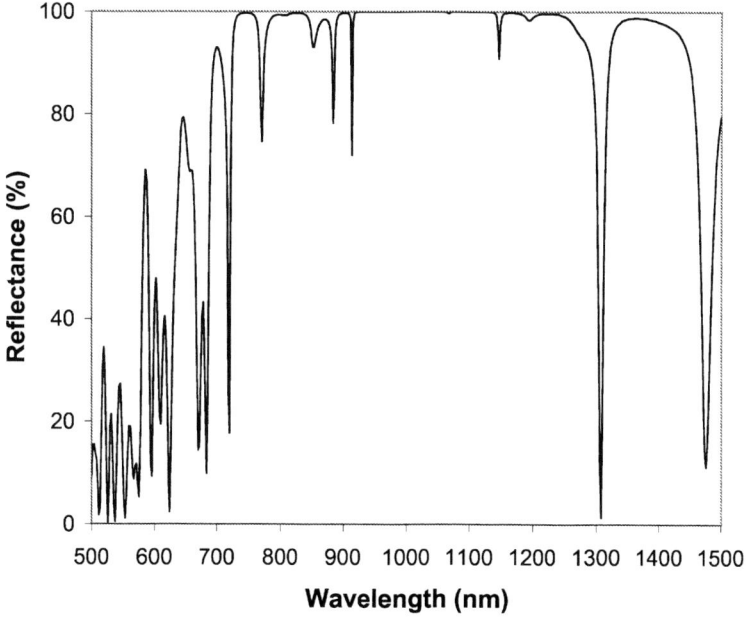

Figure 4.8 Spectral reflectance of the C-TMD described in Fig. 4.7 with the following parameters: $k = 0.25$, $f_b = 0.001$, $f_{b'} = 0.1$, and $L = 50$. The refractive indices for the ambient, substrate, and two films are the same as stated in Fig. 4.7.

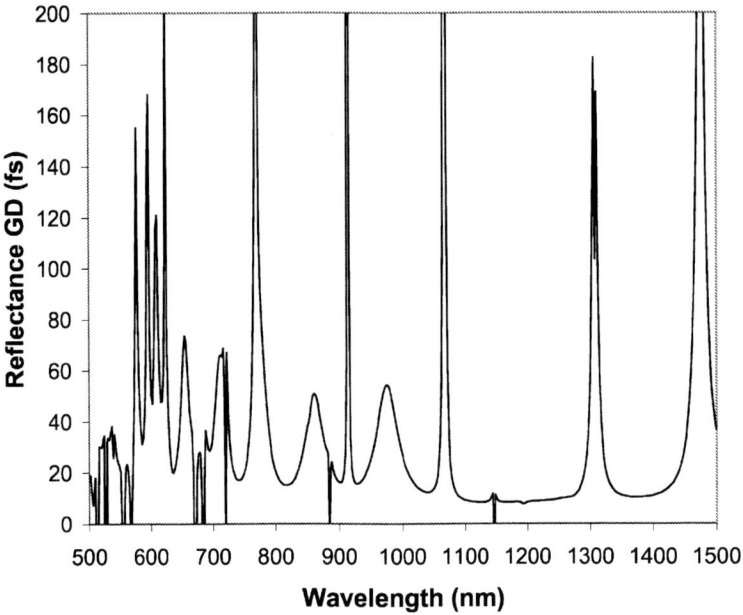

Figure 4.9 Reflectance GD of the C-TMD described in Fig. 4.8.

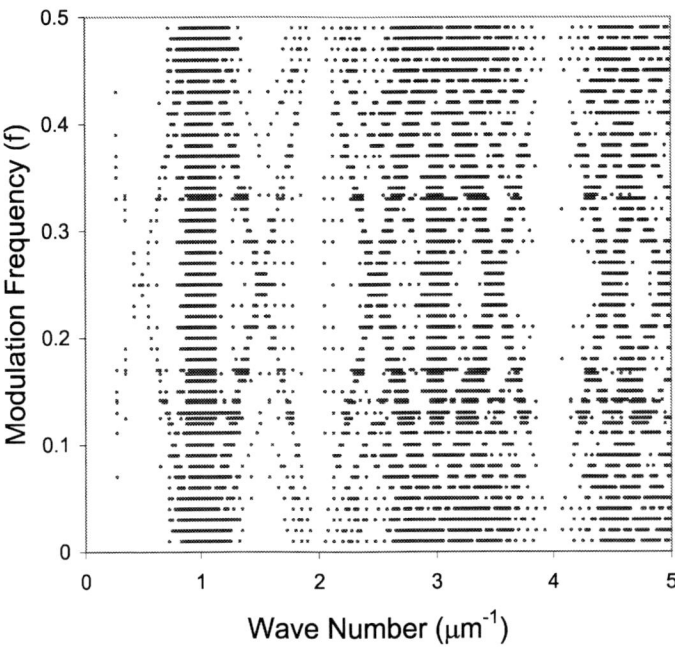

Figure 4.10 Spectral positions of all H-TMD stopbands for a modulation amplitude of 0.25. For all designs, the refractive indices for the ambient, substrate, and two films are 1.0, 1.52, 1.46, and 2.25, respectively.

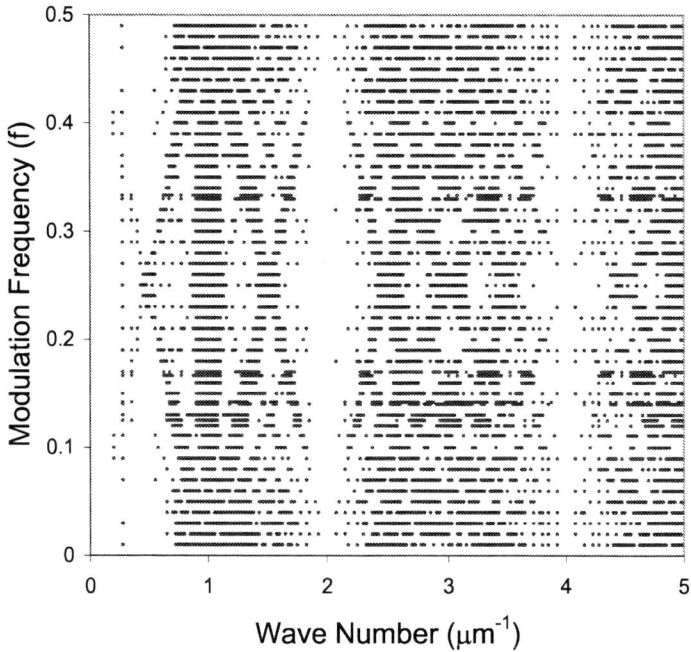

Figure 4.11 Spectral positions of all H-TMD stopbands for a modulation amplitude of 0.5. For all designs, the refractive indices for the ambient, substrate, and two films are 1.0, 1.52, 1.46, and 2.25, respectively.

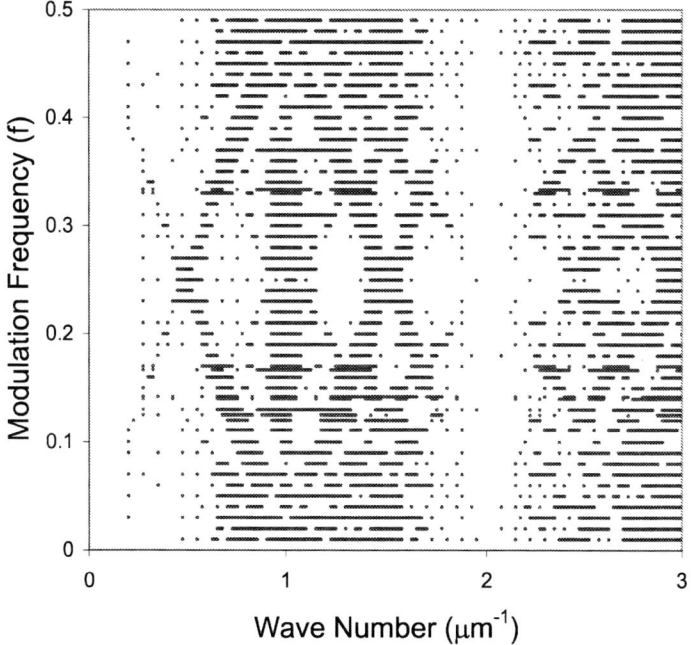

Figure 4.12 Spectral positions of all H-TMD stopbands for a modulation amplitude of 0.75. For all designs, the refractive indices for the ambient, substrate, and two films are 1.0, 1.52, 1.46, and 2.25, respectively.

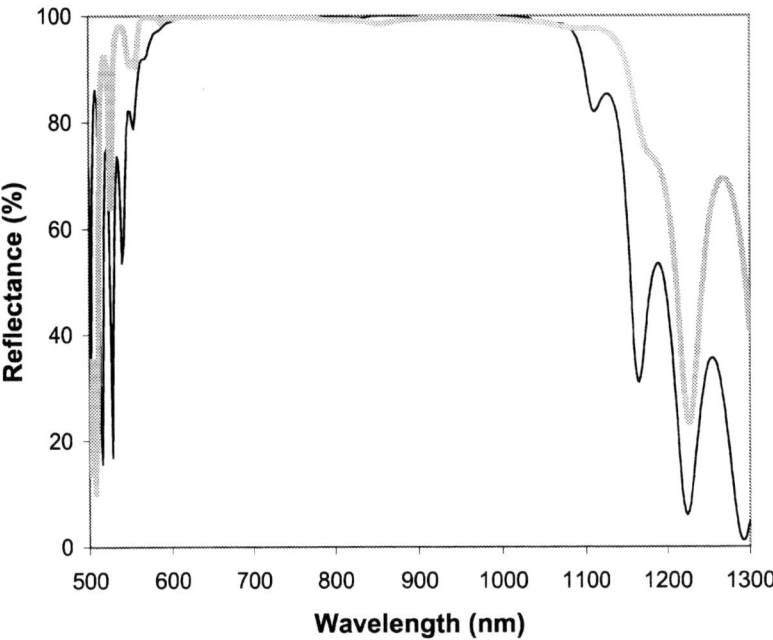

Figure 4.13 Spectral reflectance (thin black line) of the H-TMD described in Fig. 4.11 with the following parameters: $k = 0.5$, $f = 0.01$, and $L = 50$. The refractive indices for the ambient, substrate, and two films are the same as those stated in Fig. 4.7. The reflectance of the optimized design is also shown (thick gray line).

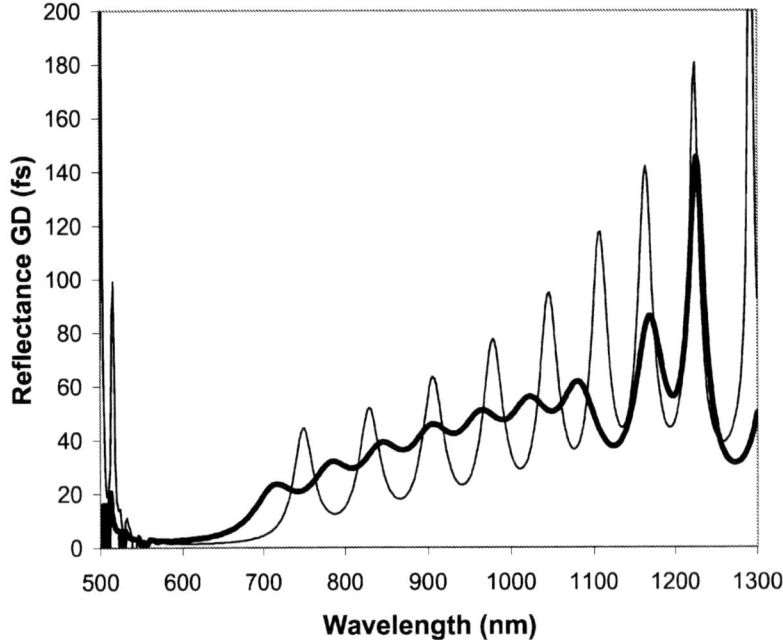

Figure 4.14 Reflectance GD of the H-TMD described in Fig. 4.13 for the unoptimized design (thin line) and optimized design (thick line). Group delay performance was optimized over the wavelength range of 600–1100 nm.

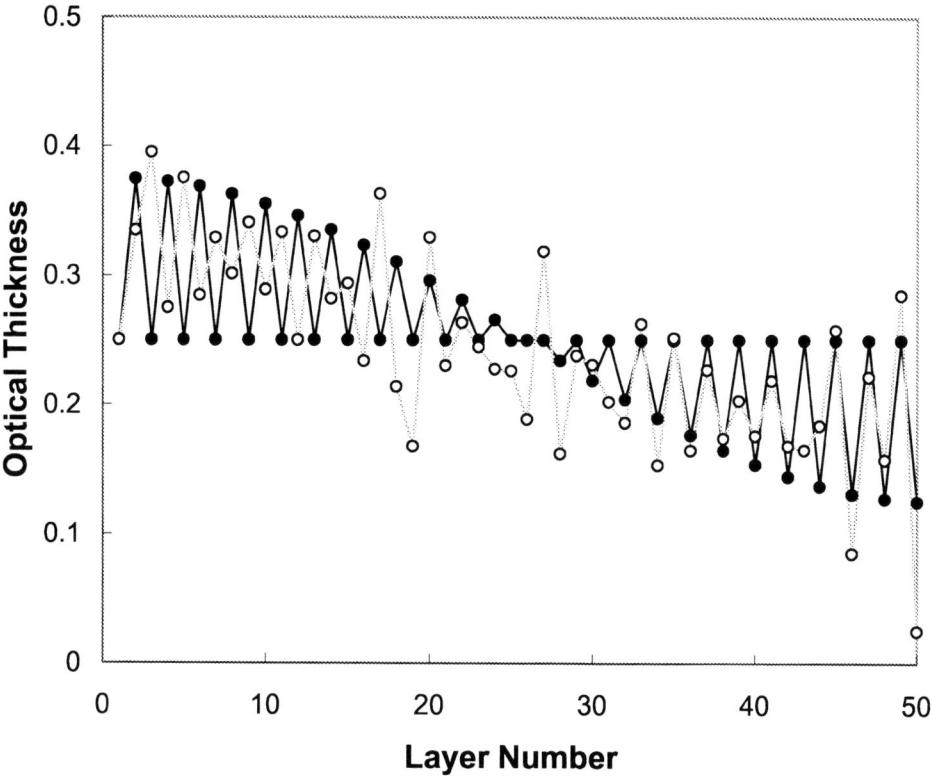

Figure 4.15 A comparison of the layer thickness for the H-TMD (black line) described in Fig. 4.13 and the optimized design (gray line) for GD.

Quarter-wave Stack Transformation

Various designs and properties of quarter-wave stacks and their respective stop-bands have been thoroughly investigated. Some design variations include high-order stacks, such as 2:1 or 3:1 stacks. In these cases, the ratio between the optical thickness of the low and high refractive-index layers is adjusted in integer multiples. Further investigation has evaluated detuned half-wave layer pairs.[25] For detuned half-wave designs, the design of a specified quarter-wave stack can be adjusted to reduce the thickness of the high refractive-index layers while preserving the desired spectral performance. This chapter investigates methods to adjust or transform a quarter-wave stack (layer thickness) into a prespecified second quarter-wave stack with a different center wavelength (for the first-order stopbands). Spectral properties of two quarter-wave stacks can be produced by partially transforming one quarter-wave stack to generate high or partial reflectance at two wavelengths. This novel transform method has not been previously published.

For the proposed transform method, analytical (linear) equations are developed that can be used for any quarter-wave stack transform. Using these analytical equations, the partial transformation is investigated where properties of both stopbands are produced. Finally, the analytical determination of prespecified properties of both stopbands is evaluated.

5.1 Stack Transformation Method

Given a 1:1 quarter-wave stack with the low and high refractive-index layers (n_L, n_H) having equal optical thickness, a stopband is produced where the center wavelength of the first-order stopband is defined to be λ_0. A second stopband is now desired to be centered at the wavelength λ_1, where $\lambda_1 > \lambda_0$ as shown in Fig. 5.1. A second quarter-wave stack with the same total thickness (new design) could be used to produce this stopband. The following stack transform method is proposed to change or transform the first quarter-wave stack into the second quarter-wave stack of equal total thickness. This transform method requires adjustment of each layer's thickness, by either increasing or decreasing it to become the QWOT of the second stack.

Figure 5.2 is a pictorial representation of the transformation of one quarter-wave stack into a second. For the example shown, groups of three layers from the first stack are transformed to become one layer of the second stack. For the three layers at the bottom of the stack, the thickness of the H layer is reduced to 0 while both L layers are increased by half the thickness of the H layer. The result of this transformation is that the QWOT thickness of the first stack is increased by a factor of three. Hence, the corresponding stopband of the second stack is three times the wavelength of the first stack.

5.2 Quarter-wave Stack Transforms

The proposed quarter-wave stack transform produces a stopband that is at a higher wavelength than the first-order stopband of the original quarter-wave stack. In order to transform a quarter-wave stack, a set of equations must be derived. These equations are first derived for one example, a special case in which the QWOT thickness of the first stack is increased by three times.

5.2.1 Example of a 3:1 quarter-wave stack transform

For a 3:1 stack transform, three layers from the first design become one layer for the second design. Because two materials (n_L, n_H) were used, a total of six transformation equations will be required (one for each of the first six layers). Once these six equations are determined, the first quarter-wave stack with L_{TOT} layers can be transformed.

Table 5.1 shows the transformation for the first six layers. To transform the first three layers (*LHL*) into one layer (*L*), both *L* layers of the first stack must be increased 50% in thickness and the *H* layer decreased by 100%. A similar pattern is followed for the next three layers. Table 5.1 also shows these six transform equations.

The last two columns in Table 5.1 show the calculated layer thicknesses for the six layers (normalized to unity). The variable k is used to express how much the first quarter-wave stack is transformed from 0 to 1. At $k = 0$ (0% transform), the equations preserve the original quarter-wave stack. At $k = 1$ (100% transform), every three layers add together to produce a new quarter-wave stack with three times the QWOT thickness of the first.

5.3 Linear Quarter-wave Stack Transformation Method

Table 5.2 outlines the procedure to transform a quarter-wave stack as described above. To help illustrate this method, an example follows Table 5.2. Further discussion is also presented in later sections of this chapter.

Table 5.1 Stack transform equations for the 3:1 case.

Layer	$T(L)$	$k = 0$	$k = 1$
1L	$1 + 0.5k$	1	1.5
2H	$1 - k$	1	0
3L	$1 + 0.5k$	1	1.5
4H	$1 + 0.5k$	1	1.5
5L	$1 - k$	1	0
6H	$1 + 0.5k$	1	1.5

Table 5.2 Quarter-wave stack transform steps.

Step	Process	Equations/Notes
1	Calculate the ratio ρ of the stopband center wavelengths λ_1 and λ_0. Adjust this ratio for dispersion and so the reciprocal of the fractional part of ρ is an integer (see Step 3 below).	$\rho = \frac{\lambda_1}{\lambda_0}$ (5.1)
2	Determine how many layers in the first stack will be uniquely transformed into the second stack (basic period of the transform). Determine the lowest integer multiple M_1 of the above ratio whose product is an even integer N. N is the required number of layers in the first stack.	$M_1\rho = N$, (5.2) where $N \in$ even integer and $M_1 \geq 2$
3	Determine the number of integer fractional units U that each layer in the original quarter-wave stack is subdivided into. U is the reciprocal of the fractional part of ρ. M_2 is any positive integer.	$U = \frac{M_2}{\text{frac}(\rho)}$ (5.3) $U = 1$ if $\rho \in$ integer
4	Determine the number of additive and subtractive fractional units U_f for each contiguous group of fractional units with the same refractive index for each quarter-wave layer in the basic period of the second stack.	See example below and Eqs. (5.5) to (5.7)
5	Determine the total number of additive and subtractive fractional units, S^+ and S^-, for each quarter-wave layer in the basic period of the second stack.	See example below and Eqs. (5.5) to (5.7)
6	Complete a table (Table 5.4) that defines the transform equations for each fractional layer in the basic period of the first quarter-wave stack.	See example below and Eqs. (5.5) to (5.7)
7	Select a percent transform k and calculate the thickness of each fractional unit.	See example below and Eqs. (5.5) to (5.7)
8	Add the calculated thickness of each fractional unit of each layer to determine the new thickness of each layer.	See example below and Eqs. (5.5) to (5.7)

5.3.1 Example of stack transform method

From Table 5.2, we first calculate ρ. For this example, an arbitrary value of 2.25 is used (Step 1). Next, Step 2 determines the lowest even integer value of N as shown in Table 5.3. Remember that M_1 must be greater than or equal to 2. The lowest positive integer value of M_2 is selected so that U is an integer or very close to an integer value (see Sec. 5.5.2 for an example of this).

Table 5.3 Direct test method to determine the lowest even integer value of N.

M_1	$M_1\rho = N$	Even Integer?
2	4.5	No
3	6.75	No
4	9	No
5	11.25	No
6	13.5	No
7	15.75	No
8	**18**	**Yes**

Therefore, as determined in Table 5.3, the first 18 layers ($N = 18$) of the original quarter-wave stack will be uniquely transformed into eight layers in the second quarter-wave stack (M_1).

Next, for Step 3, the number of fractional units is calculated from Eq. (5.3), where

$$U = \frac{M_2}{\text{frac}(\rho)} = \frac{1}{0.25} = 4. \tag{5.4}$$

The number of fractional units U is the minimum integer number of fractional sublayers that each quarter-wave layer in the first stack must be split into. Because of this requirement, the spectral centering of the second stack can be off from the target wavelength of λ_1. This depends on how close the reciprocal of the fractional part of ρ is to an integer value. The corresponding error can be reduced in some cases by increasing the value of U. Returning to the example, one transform equation will be determined for different groups of these sublayers, depending on the rules defined in Steps 4 and 5 above.

Table 5.4 shows the values of U, U_f, S^+, and S^- based on the rules of Steps 4 and 5 for the transform process. Here, there are nine fractional units per quarter-wave layer in the second stack. Starting at the first layer of the first stack, this layer and the second layer make up part of the first layer of the second stack. However, only one fractional unit of the third layer (first stack) is part of the first layer of the second stack. Hence, layer 3L is split into 1 and 3 fractional units. This process is repeated until all 18 layers of the first stack are split into fractional units that will comprise each of the eight layers of the second stack.

The transformation equations shown in Table 5.4 were derived based on the simple transform equations shown in Table 5.1. They were also derived with *a priori* knowledge that the sum of the additive and subtractive equations for each layer of the second quarter-wave stack must be combined as follows:

- The sum of the **additive** equations for each layer of the second quarter-wave stack must equal the total number of fractional units in each layer (same for all layers); and
- The sum of the **additive** and **subtractive** equations for each layer of the second quarter-wave stack must equal the total number of fractional units in each layer (same for all layers).

Table 5.4 Values of U_f, S^+ and S^-, and transform equations.

Layer 1st Stack	U	U_f	Layer 2nd Stack	S^+	S^-	± Unit	Subtractive Equations	Additive Equations
1L	4	4				+		$\frac{4}{4}(1+\frac{4}{5}k)$
2H	4	4	IL	5	4	−	$\frac{4}{4}(1-k)$	
3L	4	1				+		$\frac{1}{4}(1+\frac{4}{5}k)$
		3	2H	4	5	−	$\frac{3}{4}(1-k)$	
4H	4	4				+		$\frac{4}{4}(1+\frac{5}{4}k)$
5L	4	2				−	$\frac{2}{4}(1-k)$	
		2	3L	5	4	+		$\frac{2}{4}(1+\frac{4}{5}k)$
6H	4	4				−	$\frac{4}{4}(1-k)$	
7L	4	3				+		$\frac{3}{4}(1+\frac{4}{5}k)$
		1				−	$\frac{1}{4}(1-k)$	
8H	4	4	4H	4	5	+		$\frac{4}{4}(1+\frac{5}{4}k)$
9L	4	4				−	$\frac{4}{4}(1-k)$	
10H	4	4				−	$\frac{4}{4}(1-k)$	
11L	4	4	5L	4	5	+		$\frac{4}{4}(1+\frac{5}{4}k)$
12H	4	1				−	$\frac{1}{4}(1-k)$	
		3				+		$\frac{3}{4}(1+\frac{4}{5}k)$
13L	4	4	6H	5	4	−	$\frac{4}{4}(1-k)$	
14H	4	2				+		$\frac{2}{4}(1+\frac{4}{5}k)$
		2				−	$\frac{2}{4}(1-k)$	
15L	4	4	7L	4	5	+		$\frac{4}{4}(1+\frac{5}{4}k)$
16H	4	3				−	$\frac{3}{4}(1-k)$	
		1				+		$\frac{1}{4}(1+\frac{4}{5}k)$
17L	4	4	8H	5	4	−	$\frac{4}{4}(1-k)$	
18H	4	4				+		$\frac{4}{4}(1+\frac{4}{5}k)$

For several different examples of transforms, a general set of transform equations have been developed for additive and subtractive fractional units of layers. These equations describe this linear optical stack transform (LOST) and are given below in Eqs. (5.5) and (5.6):

$$T_{U_f}^- = \frac{T_{QW1}U_f}{U}(1-k) \qquad (5.5)$$

and

$$T_{U_f}^+ = \frac{T_{QW1}U_f}{U}\left(1 + \frac{S^-}{S^+}k\right).\tag{5.6}$$

Equation (5.5) is used for all subtractive layers and subtractive fractional parts (units) of layers. Equation (5.6) is used for additive layers and fractional units. However, Eq. (5.6) depends on the number of additive and subtractive units. A summary of all TMD equations from previous chapters and the LOST equations are tabulated in Appendix E for reference.

The additive and subtractive equations in Table 5.4 are now used to calculate the layer thickness for each of the 18 layers for a 50% stack transform, where k equals 0.5. A 50% stack transformation was selected in order to realize properties of both stopbands. Table 5.5 shows the calculated thickness from each equation

Table 5.5 Calculated thickness of each fractional unit and layer for $k = 0.5$.

Fractional Unit	Thickness of Fractional Unit
1L	1.4
2H	0.5
3L	0.35
3L	0.375
4H	1.625
5L	0.25
5L	0.7
6H	0.5
7L	1.05
7L	0.125
8H	1.625
9L	0.5
10H	0.5
11L	1.625
12H	0.125
12H	1.05
13L	0.5
14H	0.7
14H	0.25
15L	1.625
16H	0.375
16H	0.35
17L	0.5
18H	1.4

$\longrightarrow$

Transformed Layers	New Layer Thickness
1L	1.4
2H	0.5
3L	0.725
4H	1.625
5L	0.95
6H	0.5
7L	1.175
8H	1.625
9L	0.5
10H	0.5
11L	1.625
12H	1.175
13L	0.5
14H	0.95
15L	1.625
16H	0.725
17L	0.5
18H	1.4
Total	18

for fractional units, and the combined thickness for each of the 18 layers in the example design.

Note that the total thickness of the transformed layers in Table 5.5 equals the thickness of the original quarter-wave stack. This is true for all stack transformations and is independent of k.

Figure 5.3 shows the layer-thickness profile for this design, where the QWOT thickness of the original stack is normalized to unity. The spectral performances of the original stack ($k = 0$), transformed stack ($k = 0.5$), and a 100% transformed stack ($k = 1$) are shown in Figs. 5.4(a) to (c). Figure 5.5 shows the overlay of all three spectral curves. Note that as the first quarter-wave stack is transformed, the higher-order stopbands of the second stack are produced. Therefore, a new design method can be developed that uses these higher-order stopbands (see Sec. 5.4).

Figure 5.6 shows the reflectance of the center wavelength of each stopband as a function of the percent transform (k) for the above example. The reflectance of the original stack is preserved near unity until approximately a 50% transform is reached. Because the original stack is transformed into a new stack with only eight layers, or four periods [i.e., $(LH)^4$], the total reflectance is near 92% for the completed transform ($k = 1$). Additional layers in the first stack could be used to improve the reflectance of the second stopband. It is not necessary to add increments of N layers; any number of layers could be added (or deleted) from the original quarter-wave stack as long as the transform steps in Table 5.2 are followed.

5.4 High-order Harmonic Stopbands of Transformed Quarter-wave Stacks

As mentioned in the previous section, while one quarter-wave stack is transformed into another the high-order harmonic stopbands are present from both quarter-wave stacks. For example, the third-order stopband of the second quarter-wave stack is present in Fig. 5.4(c), where the original design was transformed completely (i.e., $k = 1$). The second quarter-wave stack, centered at 2.25 μm, has its third-order stopband centered at one-third this wavelength (750 nm). This transformed design (see Sec. 5.3.1) still has this stopband present, even at a 50% transform as shown in Fig. 5.4(b). Without any transformation ($k = 0$), the third-order stopband is not present, as shown in Fig. 5.4(a).

5.5 Applications

5.5.1 Dual-wavelength high-reflector example no. 1

A hypothetical design is required to produce high reflectance ($>98\%$) at $\lambda_0 = 1.0$ μm and $\lambda_1 = 1.5$ μm. Following the steps for a LOST in Table 5.2, the values of N and U are readily calculated to be 6 and 2, respectively. Therefore, every six layers of the original quarter-wave stack are uniquely transformed. This basic period of six layers can be repeated to increase the reflectance at both of the desired wavelengths. Also, the percent transform k can be varied to adjust the reflectance.

Next, Table 5.6 determines the individual LOST equations.

The reflectance of a quarter-wave stack can be calculated from the equation in Appendix A. Here, the reflectance is calculated for the wavelength at the center of the stopband. Assuming refractive indices of 1.0, 1.52, 1.45, and 2.25 for the ambient, substrate, and two thin-film layers, respectively, the reflectance was calculated as the number of two-layer periods in Fig. 5.7. For reflectance greater than 98%, at least six periods are required (12 layers total). Also, the complete transform of 12 layers (i.e., $k = 1$) will produce a new quarter-wave stack with only four periods or eight layers. Therefore, for both stacks to have greater than 98% reflectance, the minimum number of periods required is six (minimum for >98% reflectance) times ρ (see Table 5.2), or nine periods. However, the stopband reflectance decreases near $k = 0.5$ so additional periods are required. By direct calculation, the minimum number of periods to achieve >98% reflectance is 13 (26 layers). Figure 5.8 shows the reflectance at the center wavelength of each stopband as a function of k. Between approximately 0.5 and 0.6 for k, both stopbands have reflectance above 98%; also, the reflectances are equal at $k \approx 0.54$. Therefore, the selected design has 26 layers and $k = 0.54$. Lastly, the reflectance of the transformed design is shown in Fig. 5.9. The reflectance at 1000 nm and 1500 nm is 99.3% and 99.1%, respectively.

5.5.2 Dual-wavelength high-reflector example no. 2

A hypothetical design is required to produce high reflectance (>98%) at $\lambda_0 = 1064$ nm and $\lambda_1 = 633$ nm. Also, the bandwidth of the stopband at 633 nm must be minimized.

Using the design method presented in Sec. 5.4, the third-order stopband of the transformed design can be used to reflect 633 nm. This design method takes advantage of the smaller bandwidth of the third-order stopband. Therefore, for

Table 5.6 Values of U_f, S^+ and S^-, and transform equations.

Layer 1st Stack	U	U_f	Layer 2nd Stack	S^+	S^-	$\pm$ Unit	Subtractive Equations	Additive Equations
1L	2	2				$+$		$\frac{2}{3}(1 + \frac{1}{2}k)$
2H	2	1	1L	2	1	$-$	$\frac{1}{2}(1 - k)$	
		1	2H	1	2	$+$		$\frac{1}{2}(1 + 2k)$
3L	2	2				$-$	$\frac{2}{3}(1 - k)$	
4H	2	2				$-$	$\frac{2}{3}(1 - k)$	
5L	2	1	3L	1	2	$+$		$\frac{1}{2}(1 + 2k)$
		1				$-$	$\frac{1}{2}(1 - k)$	
6H	2	2	4H	2	1	$+$		$\frac{2}{3}(1 + \frac{1}{2}k)$

this transformation, λ_1 will be three times 633 nm, or 1899 nm. Accordingly, ρ is calculated to be 1.784. By direct testing of M_1 values, the number of layers that are uniquely transformed is very close to 25 (24.976) when $M_1 = 14$. Because the number of layers is not **exactly** an integer, the last layer's thickness (layer 25) will transform into a new thickness that is slightly off from the other layers (this is explained below).

Continuing with Step 3 from Table 5.2, U is determined by direct testing of M_2 values starting at 1. Here, when $M_2 = 29$, $U = 36.989$, which is very close to the integer 37. Therefore, 37 is selected as the number of fractional units, U, that each layer's thickness will be divided into. Note that there is a maximum of three LOST equations for each transformed layer, so a large number of fractional units does not increase the number of LOST equations.

A table similar to Tables 5.4 and 5.6 has been generated for this design example that covers the first 25 layers of this design. The unique part of this table is shown in Table 5.7. Here, the last four layers are split into fractional units that are combined to form the last three layers of the second quarter-wave stack. There are 66 fractional units for each layer 1 through 13 of the second quarter-wave stack. The last layer of the second stack, 14H, has 67 fractional units of the first stack because when $M_1 = 14$, the number of layers that is uniquely transformed is not exactly 25. However, the LOST equations account for this difference.

Figure 5.10 shows the reflectance of this design as a function of k at wavelengths of 633 nm and 1064 nm. The reflectance at both wavelengths is approximately 90% when $k \approx 0.7$. Additional layers could be added to increase the reflectance.

Table 5.7 Values of $U_{f'}$, S^+ and S^-, and transform equations.

Layer 1st Stack	U	U_f	Layer 2nd Stack	S^+	S^-	$\pm$ Unit	Subtractive Equations	Additive Equations
22H	37	15	Part of 12H	n/a	n/a	n/a	n/a	n/a
		22	13L	37	29	$-$	$\frac{22}{37}(1-k)$	
23L	37	37				$+$		$\frac{37}{37}(1+\frac{29}{37}k)$
24H	37	7	14H	30	37	$-$	$\frac{7}{37}(1-k)$	
		30				$+$		$\frac{30}{37}(1+\frac{37}{30}k)$
25L	37	37				$-$	$\frac{37}{37}(1-k)$	

5.6 Exercises

1. What is the maximum reflectance of a second stopband, where $\lambda_1 > \lambda_0$, that can be achieved for a given number of layers $(LH)^n$ and refractive indices for the original quarter-wave stack?
2. Prove that the analytical equation given below calculates the same layer thickness for a 1:3 stack transformation using LOST equations.

$$T(L) = T_{AVG}\left\{1 + k\cos\left[\frac{\pi}{3}(2L - 1)\right]\right\}.$$

3. What is the average thickness for the low and high refractive-index layers as a function of the general case of a linear stack transform? How does this affect the spectral centering of the second stopband?
4. Derive the LOST equations that would transform a 1:1 quarter-wave stack into a second quarter-wave stack that has a layer thickness ratio of 2:1 between the low and high refractive-index layers.
5. Develop a method to achieve a reverse LOST, where the desired stopband is at a shorter wavelength than the first-order stopband of the original quarter-wave stack (i.e., $\lambda_1 < \lambda_0$).
6. Calculate the reflectance of the stopband centers as a function of k for $\rho = 2$, 3, 4, and 5.
7. Calculate the bandwidth of both stopbands as a function of k and the refractive indices.
8. Write a computer program that calculates the transformed layer thickness for the general case of a LOST.

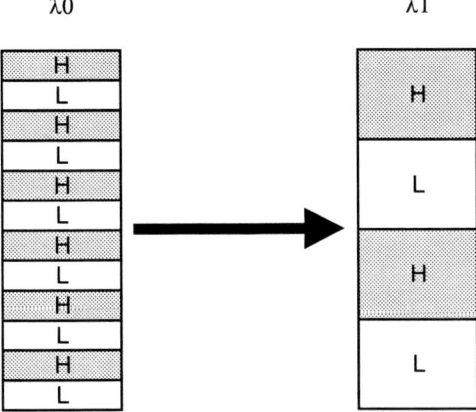

Figure 5.1 A 1:1 quarter-wave stack (left) can be transformed into a second 1:1 quarter-wave stack (right) that has thicker layers. Note that for this transformation, the total thicknes of the first stack does not change.

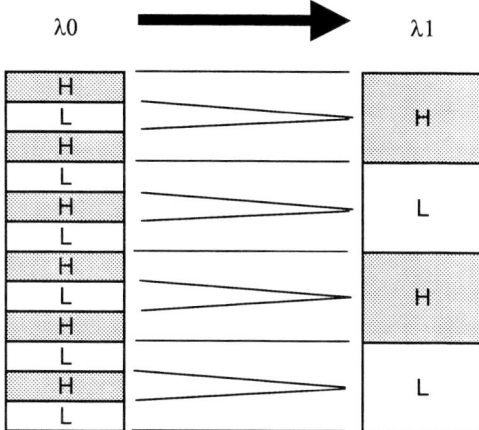

Figure 5.2 Example of a quarter-wave stack transformation. Here, the initial quarter-wave stack on the left has thickness adjustments to each layer. For this example, the thicknesses every three layers are adjusted to become one layer of the final quarter-wave stack.

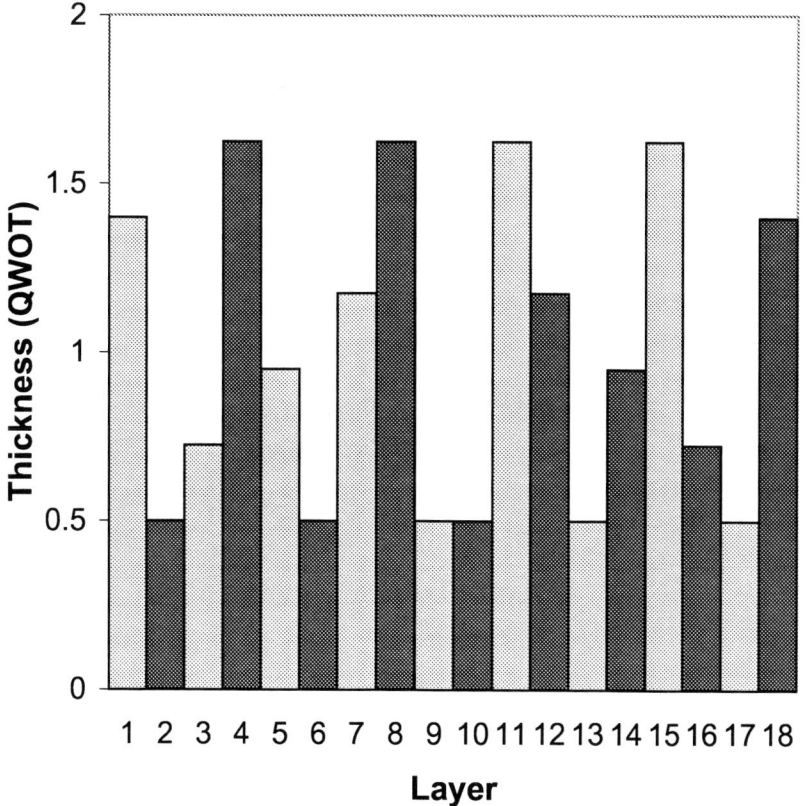

Figure 5.3 Layer-thickness profile of a 50% stack transform ($k = 0.5$) for the design example in Sec. 5.3.1.

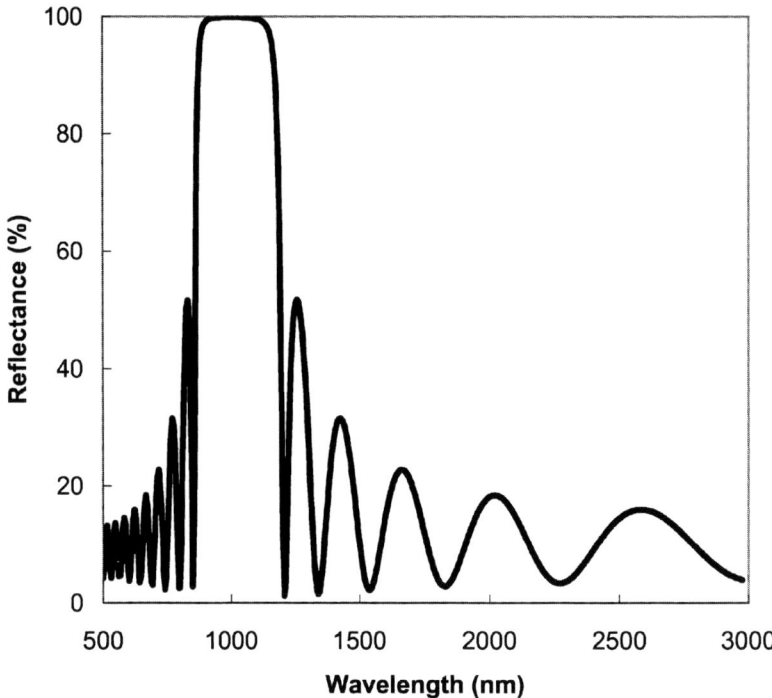

Figure 5.4(a) Reflectance of the stack transform design in Sec. 5.3.1 for a 0% transformation ($k = 0$).

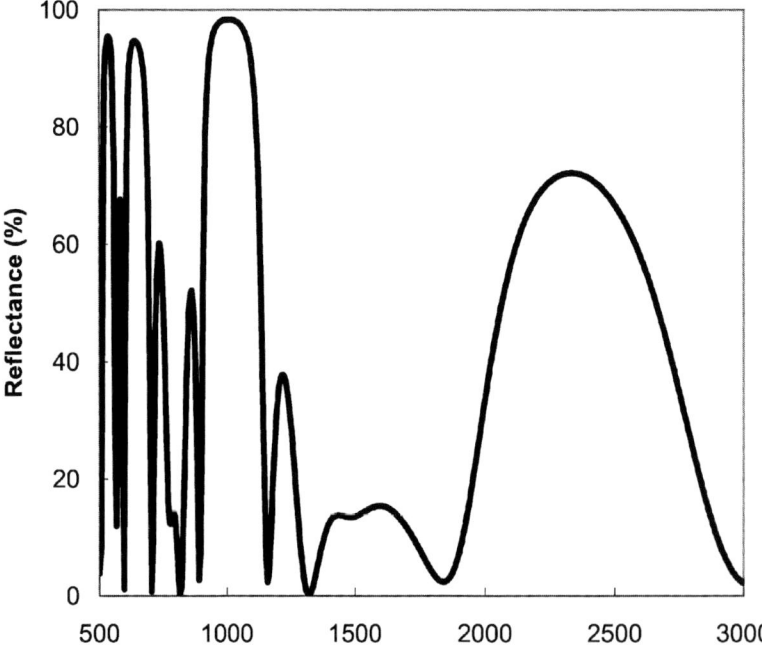

Figure 5.4(b) Reflectance of the stack transform design in Sec. 5.3.1 for a 50% transformation ($k = 0.5$).

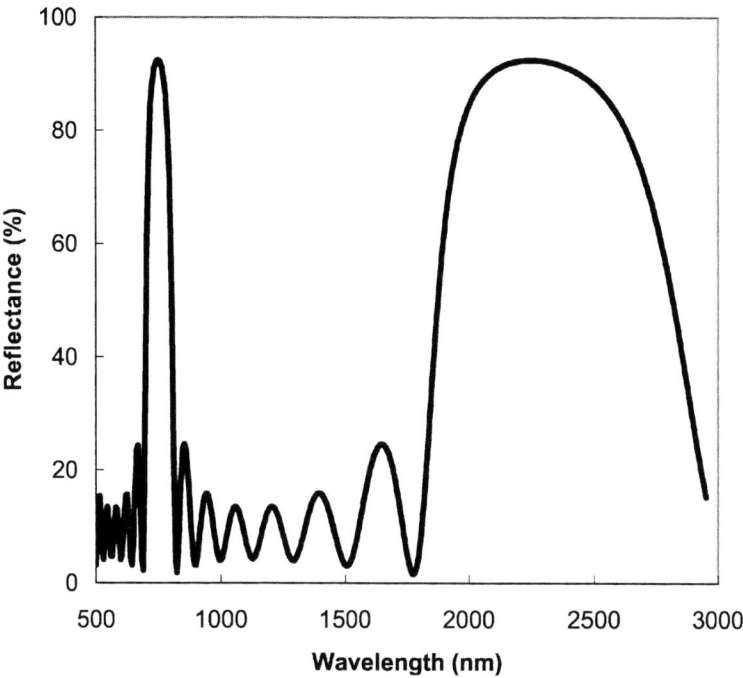

Figure 5.4(c) Reflectance of the stack transform design in Sec. 5.3.1 for a 100% transformation ($k = 1$).

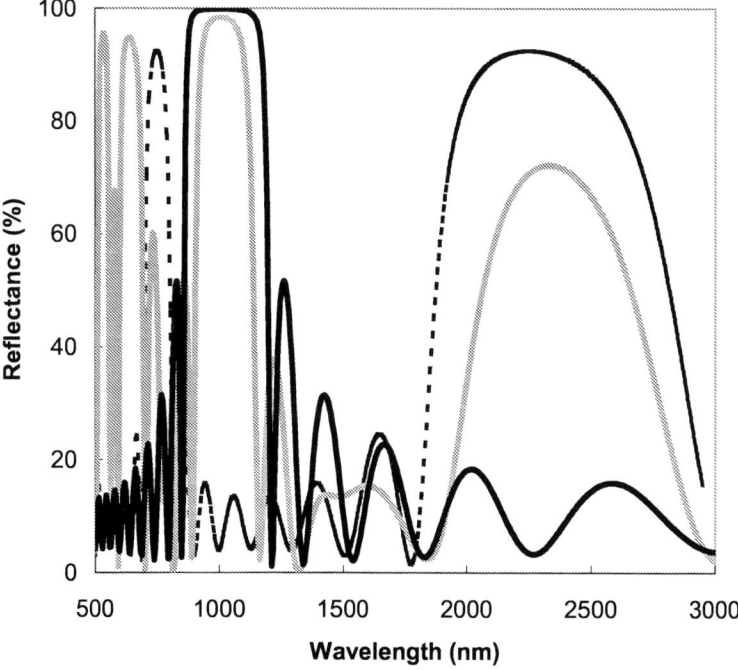

Figure 5.5 Reflectance of the stack transform design in Sec. 5.3.1 for 0% (black line, $k = 0$), 50% (gray line, $k = 0.5$), and 100% (dashed, $k = 1.0$) transformations.

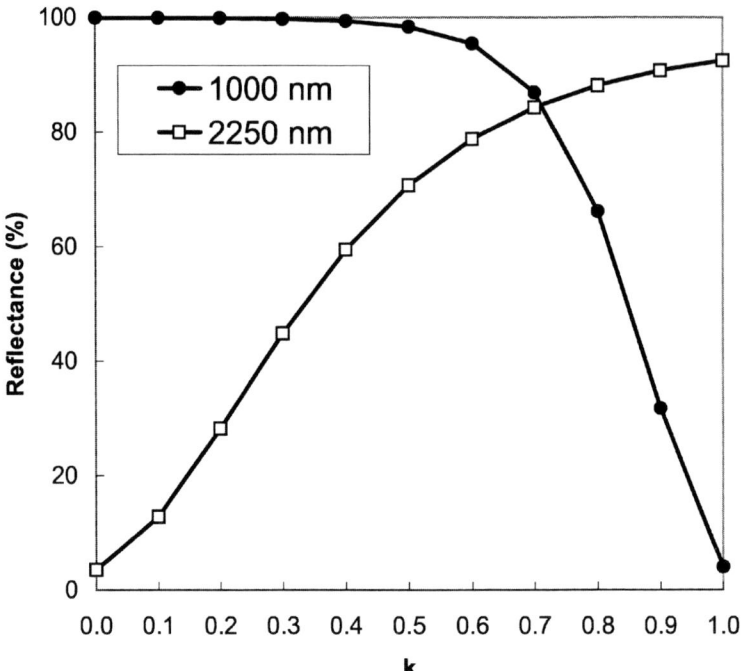

Figure 5.6 Reflectance of the stack transform design in Sec. 5.3.1 for the center wavelength of each stopband as a function of the percentage of the transform (k).

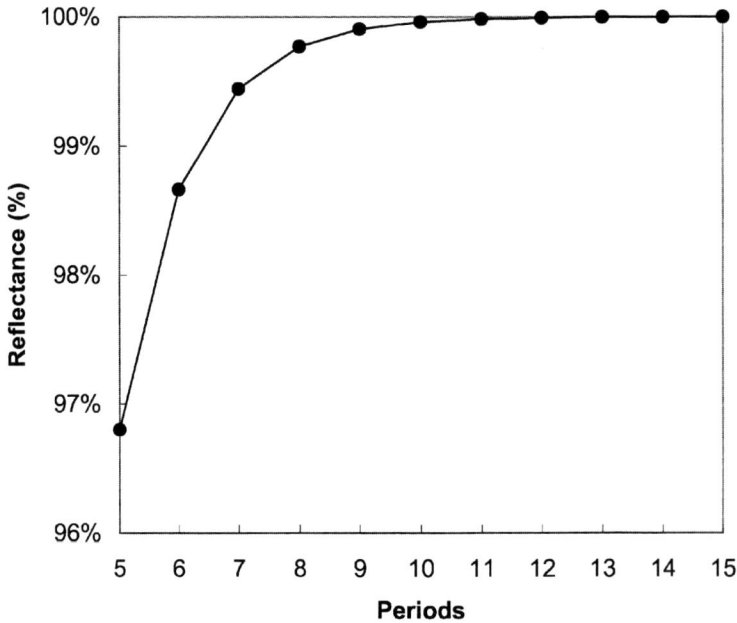

Figure 5.7 Calculated reflectance of a quarter-wave stack at the center wavelength of the first-order stopband at normal incidence. The ambient, substrate, and two films' refractive indices are 1.0, 1.52, 1.45, and 2.25, respectively. (See equation in Appendix A.)

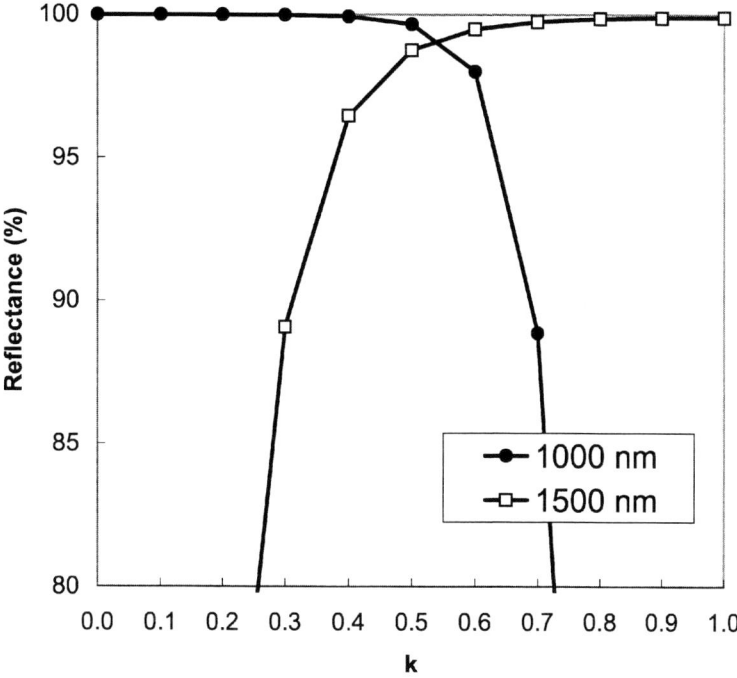

Figure 5.8 Reflectance of the stack transform design in Sec. 5.5.1 for the center wavelength of each stopband as a function of the amount of transform (k).

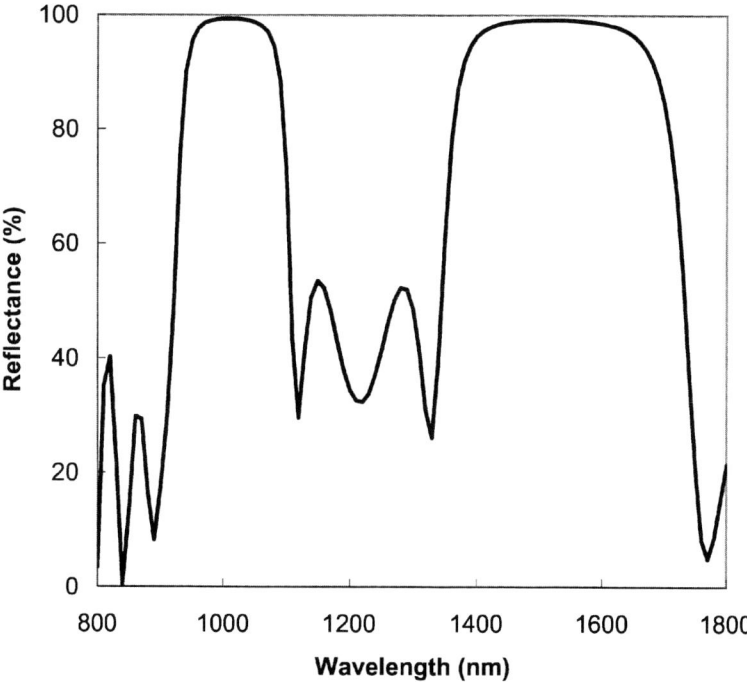

Figure 5.9 Reflectance of the stack transform design in Sec. 5.5.1 for a 54% transformation ($k = 0.54$).

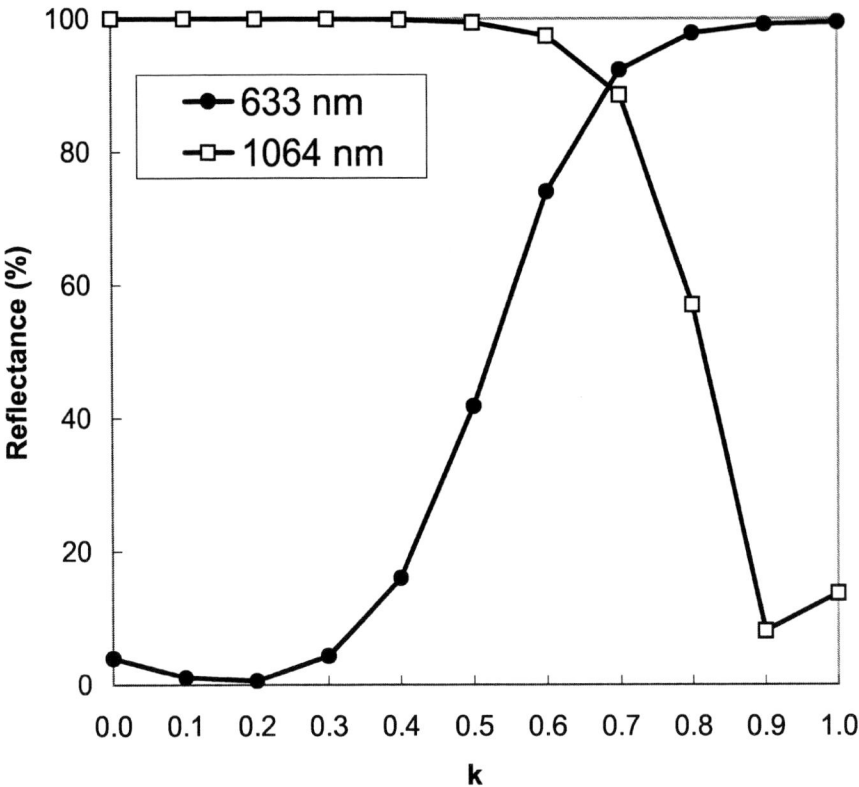

Figure 5.10 Reflectance of the stack transform design in Sec. 5.5.2 for the center wavelength of each stopband as a function of the amount of transform (k).

Useful Equations for Discrete-layer Thin-Film Calculations

	For uncoated nonabsorbing substrates	For integer-multiples of quarter-wave thickness of a nonabsorbing film or nonabsorbing substrates
To calculate refractive index	For $n_s \geq 1$: $n_s = \dfrac{1 + \sqrt{R}}{1 - \sqrt{R}}$ For $n_s < 1$: $n_s = \dfrac{1 - \sqrt{R}}{1 + \sqrt{R}}$	For $n_f \geq \sqrt{n_s}$: $n_f = \sqrt{n_s \left[\dfrac{1 + \sqrt{R}}{1 - \sqrt{R}}\right]}$ For $n_f < \sqrt{n_s}$: $n_f = \sqrt{n_s \left[\dfrac{1 - \sqrt{R}}{1 + \sqrt{R}}\right]}$
To calculate reflectance	$R_s = \left(\dfrac{n_s - 1}{n_s + 1}\right)^2$	$R_f = \left(\dfrac{1 - \dfrac{n_f^2}{n_s}}{1 + \dfrac{n_f^2}{n_s}}\right)^2$
To calculate the center wavelength, stopband reflectance for an N-period quarter-wave stack at normal incidence	$R = \left[\dfrac{1 - \dfrac{n_s}{n_0}\left(\dfrac{n_H}{n_L}\right)^{2N}}{1 + \dfrac{n_s}{n_0}\left(\dfrac{n_H}{n_L}\right)^{2N}}\right]^2$	

$\dfrac{\lambda}{4} =$

Chebyshev Polynomials of the Second Kind

The table in this appendix lists the first seven orders of Chebyshev polynomials of the second kind as a function of the argument a (see Sec. 2.1.1). The first six orders shown are from Ref. [2].

Order	Chebyshev Polynomial
$C_0(a)$	1
$C_1(a)$	$2a$
$C_2(a)$	$4a^2 - 1$
$C_3(a)$	$8a^3 - 4a$
$C_4(a)$	$16a^4 - 12a^2 + 1$
$C_5(a)$	$32a^5 - 32a^3 + 6a$
$C_6(a)$	$64a^6 - 80a^4 + 24a^2 - 1$

FORTRAN90 Source Code for the Determination of All Possible TMD Stopbands

```
!    *******************START OF MAIN PROGRAM**********
     PROGRAM TMDTRACE

!    PROGRAM:    TMDTRACE.F90
!    WRITTEN BY:  BRUCE PERILLOUX
!
!    NOTE: THIS PROGRAM IS PROVIDED FOR REFERENCE ONLY.
!    SUPPORT WILL NOT BE PROVIDED FOR THIS CODE.
!
!
!    **************************************************
!    DESCRIPTION:
!    THIS PROGRAM ACCEPTS INPUT FROM A USER-DEFINED
!    TXT FILE, CALCULATES LAYER THICKNESS, AND COMPUTES
!    THE TRACE OF THE CHARACTERISTIC MATRIX FOR THICKNESS
!    MODULATED THIN-FILM DESIGNS.
!    NEXT, THE PROGRAM DETERMINES THE SPECTRAL ENDPOINTS
!    OF ALL POSSIBLE STOPBANDS WHERE THE TRACE OF THE
!    CHARACTERISTIC MATRIX IS ABOVE 2. FOR EACH PRESPECIFIED
!    MODULATION FREQUENCY (THIN-FILM DESIGN),
!    THE PROGRAM DETERMINES THE WAVELENGTHS WHERE THE
!    TRACE IS ABOVE THE THRESHOLD VALUE FOR A STOPBAND.
!    LAST, THE PROGRAM SAVES THE START-STOP WAVELENGTHS
!    FOR EACH STOPBAND SEQUENTIALLY, FOR EACH DESIGN
!    IN AN ASCII TEXT OUTPUT FILE.  THE OUTPUT DATA IS
!    FORMATTED SO THE STOPBANDS CAN BE GRAPHED BY MOST
!    COMMERCIALLY AVAILABLE GRAPHICS PROGRAMS AS A
!    FUNCTION OF MODULATION FREQUENCY.
!    **************************************************
```

```
!    ******************************************************
!    NOTES ON INPUT DATA
!    THE PROGRAM LIMITS THE NUMBER OF INPUT MODULATION
!    FREQUENCIES TO 500.
!    THE MAXIMUM NUMBER OF LAYERS ACCEPTED FOR ANY TMD
!    IS 200. THE TOTAL NUMBER OF LAYERS FOR EACH TMD
!    IS ASSUMED TO BE AN EVEN NUMBER.
!    ******************************************************
!    VARIABLE DEFINITIONS:
!
!    FNAME - NAME OF THE INPUT DATA FILE
!    FOUT - NAME OF THE OUTPUT DATA FILE
!    F(500) - MODULATION FREQUENCY FOR EACH TMD (500
!             MAX DESIGNS)
!    INSB - FLAG TO INDICATE START/STOP OF STOP BANDS
!    J - GENERAL USE INTEGER FOR PROGRAM LOOPS
!    K - GENERAL USE INTEGER FOR PROGRAM LOOPS
!    KMOD - TMD MODULATION AMPLITUDE FOR ALL TMD'S
!    L(500) - NUMBER OF LAYERS FOR EACH TMD OR TMD
!             PERIOD
     M - LAYER NUMBER; USED TO CALCULATE TMD
!             THICKNESS
     N(200) - REFRACTIVE INDEX OF EACH LAYER,
!    NH, NL (LAYER 1 = NL)
!      ***NOTE: LAYER 1 IS ADJACENT TO THE AMBIENT
!             MEDIUM
     N0 - INDEX OF REFRACTION FOR AMBIENT MEDIUM
!    (REAL)
     NL - INDEX OF REFRACTION FOR LOW-INDEX
!    LAYER MATERIAL (REAL)
     NH - INDEX OF REFRACTION FOR
!             HIGH-INDEX LAYER MATERIAL (REAL)
     NS - INDEX OF REFRACTION FOR SUBSTRATE MEDIUM (REAL)
!    PREVS - INITIALIZATION VARIABLE FOR WAVE NUMBER
     RSTART - INITIALIZATION VARIABLE FOR REFLECTANCE
     SMIN - MINIMUM WAVE NUMBER FOR STOPBAND SEARCH RANGE
!    SMAX - MAXIMUM WAVE NUMBER FOR STOPBAND SEARCH RANGE
!    SINC - WAVE NUMBER INCREMENT FOR STOPBAND SEARCH RANGE
!    SIGMA - WAVE NUMBER
     T(500,200) - OPTICAL THICKNESS (QWOT) FOR EACH TMD
!       *** NOTE: 200 LAYERS MAX. FOR EACH TMD
!    TR - LIMIT OF TRACE FOR STOPBANDS (=2)
!    TRACE - CALCULATED TRACE OF THE CHARACTERISTIC
!           MATRIX OR THE COMPLETE TMD
!    W - INTEGER NUMBER OF WAVE NUMBERS PER TMD LOOP
!    X - THE NUMBER OF TMD'S IN THE INPUT FILE (500 MAX.)
!
!    ******************************************************
```

```
!    SUBROUTINE LIST:
!
!    CALC - CALCULATES THE TRACE FOR A GIVEN TMD
!    MULT - MULTIPLIES A 2X2 MATRIX
!    **************************************************
!
!

     IMPLICIT NONE

     INTEGER J,X,K,M,W,L(500)
     REAL N0,NS,NH,NL,KMOD,F(500),N(200),T(500,200)
     REAL SMIN,SMAX,SINC,SIGMA
     REAL RSTART,TR,PREVS
     REAL TRACE
     INTEGER*2 INSB
     CHARACTER*64 FNAME,FOUT

!    DATA INPUT SECTION
     WRITE(*,'(//'' Enter the name of the input text file: ''\)')
     READ(*,'(1A64)')FNAME

     WRITE(*,'(//'' Enter the name of the OUTput text file: ''\)')
     READ(*,'(1A64)')FOUT

     OPEN(4,FILE=FNAME,STATUS='UNKNOWN')

     READ(4,*)N0
     READ(4,*)NS
     READ(4,*)NH
     READ(4,*)NL
     READ(4,*)KMOD
     READ(4,*)SMIN
     READ(4,*)SMAX
     READ(4,*)SINC
     TR = 2.0          !*** this is the trace limit
                            for HR's to exist:  2 **
     J=0
5    J=J+1
     READ(4,*,END=9)F(J),L(J) ! MOD. FREQ. AND NUMBER OF LAYERS
     GOTO 5
9    CONTINUE
     CLOSE(4,STATUS='KEEP')

!    INPUT FILE IS CLOSED

     X=J-1  !THIS IS THE NUMBER OF TMD DESIGNS
!    DATA INPUT IS COMPLETE
```

```
!    START PRECALCULATIONS

!    FIRST, SET UP THE MATRIX FOR ALTERNATING REFRACTIVE INDICES.
!    THE FIRST LAYER IS ADJACENT TO AMBIENT (NOT SUBSTRATE)!,
!    THE HIGH REFRACTIVE INDEX (NH) LAYER IS ADJACENT TO THE
!    SUBSTRATE (ALL TMDS ARE ASSUMED TO HAVE AN EVEN NUMBER
!    OF LAYERS).

     DO J=1,100
         N((2*J)-1)=NL
         N(2*J)=NH
     END DO


!    SECOND, SET UP THE MATRIX FOR LAYER THICKNESS.
!    REMEMBER, THE OUTER LAYER IS T(J,1)
     DO J=1,X      !NUMBER OF TMD DESIGNS
!    WRITE(*,*)F(J),L(J)
!    PAUSE
         DO K=1,L(J)   !NUMBER OF LAYERS IN EACH TMD DESIGN
             M=L(J)-K+1
             T(J,K)=1+(KMOD*COSD(360.0*F(J)*M))
!            WRITE(*,*)K,M,T(J,K),N(K)
             T(J,K)=T(J,K)/(4.0*N(K))
             ! THICKNESS IS QWOT=1 MICROMETER (WAVELENGTH)
             ! OR ONE WAVE NUMBER (MICROMETER-1)
         END DO
!        PAUSE
     END DO
!    PRECALCULATIONS ARE COMPLETE

!    PAUSE 'STARTING ****TRACE******* CALCULATIONS'

!    NOW CALCULATE TRACE

     OPEN(4,FILE=FOUT,STATUS='UNKNOWN')

     DO K=1,X         !NUMBER OF TMD DESIGNS
         WRITE(*,'(1X,1I4,'' OF '',1I4)')K,X
         W=1+INT(((SMAX-SMIN)/SINC)+.01)
!        WRITE(*,*)SMIN,SMAX,SINC,W
!        PAUSE
         PREVS=0.0    !INITIALIZE STORAGE VAR. FOR SIGMA
         RSTART=0.0   !INITIALIZE REFLECTANCE
         INSB=0       !INITIALIZE STOPBAND FLAG
         DO J=1,W     !SCAN WAVE NUMBERS
                 SIGMA=SMIN+(SINC*REAL(J-1))
                 CALL CALC(SIGMA,L(K),N,T,K,TRACE)
```

```fortran
                        IF(TRACE .GT. TR)THEN
                            PREVS=SIGMA
                            IF(INSB .EQ. 0)THEN
                            !FOUND START OF STOPBAND   (TRACE > 2)
                            IF(J .NE. W)WRITE(4,10)SIGMA,F(K)
 10       FORMAT(2(1X,1F8.5))
                                INSB=1         !FLAG INSIDE STOPBAND
                        ELSE
                            IF(J .EQ. W)THEN
                                    WRITE(4,10)SIGMA,F(K)
                            WRITE(4,*)
                        ENDIF
                        ENDIF
                        ELSE
                          IF(INSB .EQ. 1)THEN
                            ! FOUND END OF STOPBAND
                            WRITE(4,10)PREVS,F(K)
                            WRITE(4,*)   !BLANK LINE FOR GRAPH
                            INSB=0  !FLAG OUTSIDE STOPBAND
                          ENDIF
                        ENDIF
          END DO
          END DO
! ALL STOPBANDS HAVE BEEN FOUND


    CLOSE(4,STATUS='KEEP')

    STOP
    END PROGRAM

    SUBROUTINE CALC(SIGMA,L,N,T,TMD,TRACE)

    IMPLICIT NONE

    INTEGER K,L,A,B,TMD
    REAL N(200),T(500,200),SIGMA,PHI,TRACE
    DOUBLE COMPLEX M(200,2,2)
    COMMON /A/ M

!   WRITE(*,'('' N  T  SIGMA'')')
    DO K=1,L
!   WRITE(*,'(1X,1I3,3(2X,F8.4))')K,N(K),T(TMD,K),SIGMA
        PHI=360.0*N(K)*T(TMD,K)*SIGMA
        M(K,1,1)=DCMPLX(COSD(PHI),0.0)
        M(K,1,2)=DCMPLX(0.0,((SIND(PHI))/N(K)))
        M(K,2,1)=DCMPLX(0.0,(N(K)*SIND(PHI)))
        M(K,2,2)=M(K,1,1)
```

```
      END DO
!     PAUSE

      IF(L .NE. 1)THEN
        DO K=1,L-1
             A=L-K
             B=A+1
             CALL MULT(A,B)
        END DO
      ENDIF

!     NOW CALCULATE THE TRACE OF THE CHARACTERISTIC MATRIX
      TRACE=ABS(REAL(M(1,1,1))+REAL(M(1,2,2)))

      RETURN
      END


      SUBROUTINE MULT(A,B)

      DOUBLE COMPLEX M(200,2,2),MS(2,2)
      INTEGER A,B

      COMMON /A/ M

      MS(1,1)=(M(A,1,1)*M(B,1,1))+(M(A,1,2)*M(B,2,1))
      MS(1,2)=(M(A,1,1)*M(B,1,2))+(M(A,1,2)*M(B,2,2))
      MS(2,1)=(M(A,2,1)*M(B,1,1))+(M(A,2,2)*M(B,2,1))
      MS(2,2)=(M(A,2,1)*M(B,1,2))+(M(A,2,2)*M(B,2,2))
      M(A,1,1)=MS(1,1)
      M(A,1,2)=MS(1,2)
      M(A,2,1)=MS(2,1)
      M(A,2,2)=MS(2,2)

      RETURN
      END
```

Stopband Positions for Selected TMDs

The spectral stopband positions of three groups of TMDs are shown here. Each group has four plots where each group has the same refractive-index data shown in Table D1 below.

Table D1 Refractive-index data for each TMD group.

TMD Group	n(substrate)	$n(H)$	$n(L)$	Figures
1	1.46	2.0	1.45	D1, D2, D3, D4
2	2.46	2.2	1.4	D5, D6, D7, D8
3	4.0	4.0	2.2	D9, D10, D11, D12

The four figures in each TMD group above correspond to the TMD modulation amplitudes of 0.25, 0.5, 0.75, and 1.0. The modulation frequency and number of layers for each TMD in every plot are shown below.

f	L_{TOTAL}
0.01	100
0.05	40
0.02	50
0.1	30
0.1111111	36
0.125	32
0.142	28
0.166666	30
0.2	30
0.25	24
0.3333333	24
0.4	30
0.5	20
0.03	100
0.04	50
0.06	50
0.07	50
0.08	50
0.09	50
0.12	50
0.13	50

f	L_{TOTAL}
0.14	50
0.15	50
0.16	50
0.17	50
0.18	50
0.19	50
0.21	50
0.22	50
0.23	50
0.24	50
0.26	50
0.27	50
0.28	50
0.29	50
0.3	50
0.31	50
0.32	50
0.33	50
0.34	50
0.35	50
0.36	50
0.37	50
0.38	50
0.39	50
0.41	50
0.42	50
0.43	50
0.44	50
0.45	50
0.46	50
0.47	50
0.48	50
0.49	50

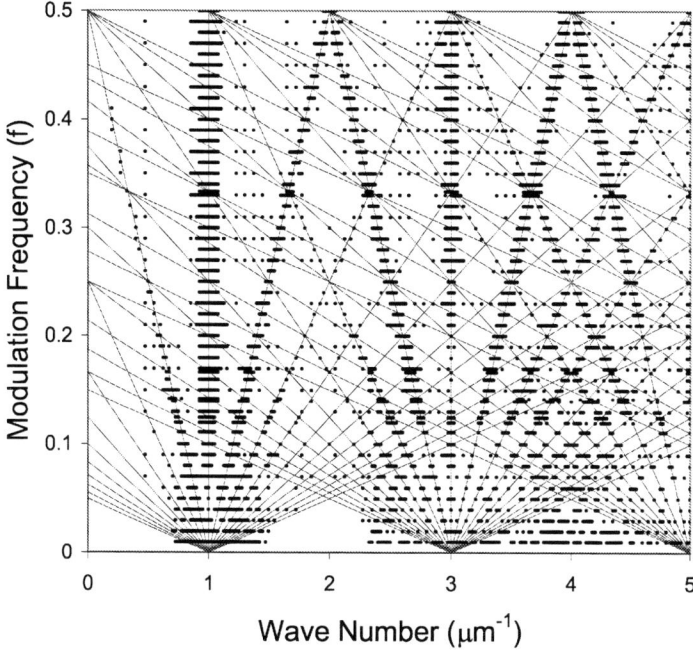

Figure D1 Spectral positions of all stopbands for a modulation amplitude of 0.25. For all designs, the refractive indices for the ambient, substrate, and two films are 1.0, 1.46, 1.45, and 2.0, respectively.

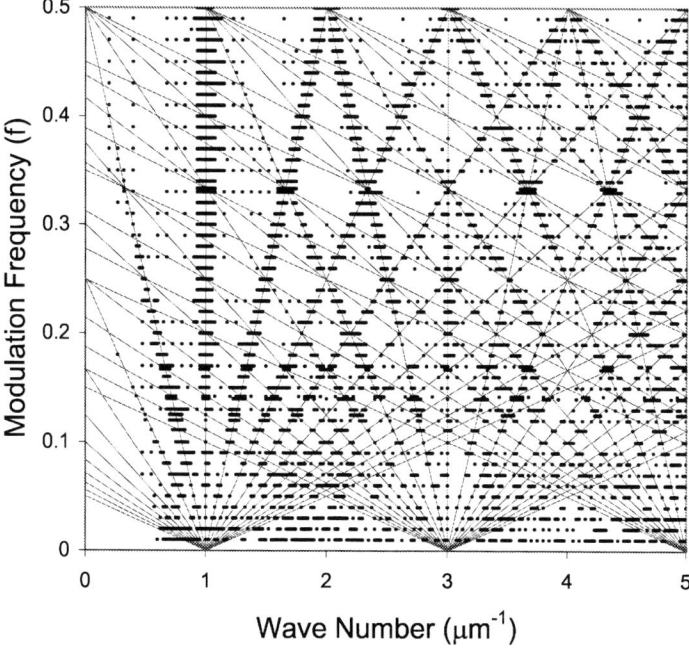

Figure D2 Spectral positions of all stopbands for a modulation amplitude of 0.5. For all designs, the refractive indices for the ambient, substrate, and two films are 1.0, 1.46, 1.45, and 2.0, respectively.

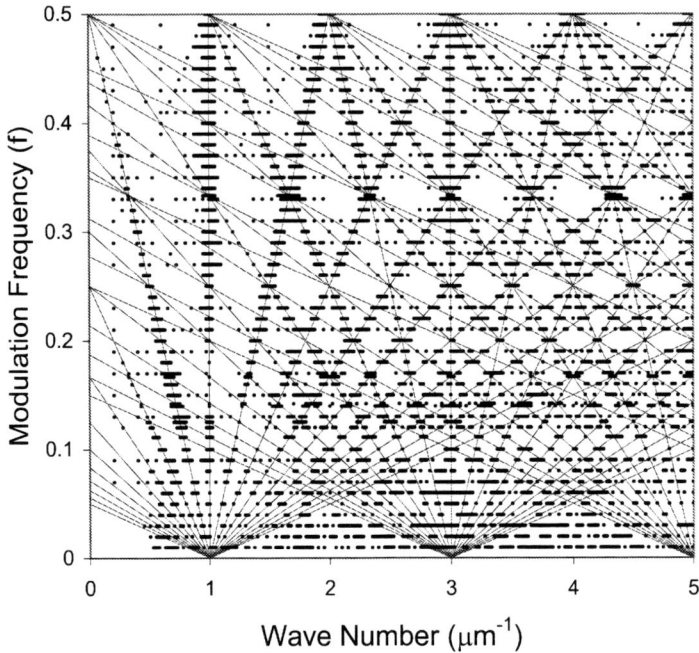

Figure D3 Spectral positions of all stopbands for a modulation amplitude of 0.75. For all designs, the refractive indices for the ambient, substrate, and two films are 1.0, 1.46, 1.45, and 2.0, respectively.

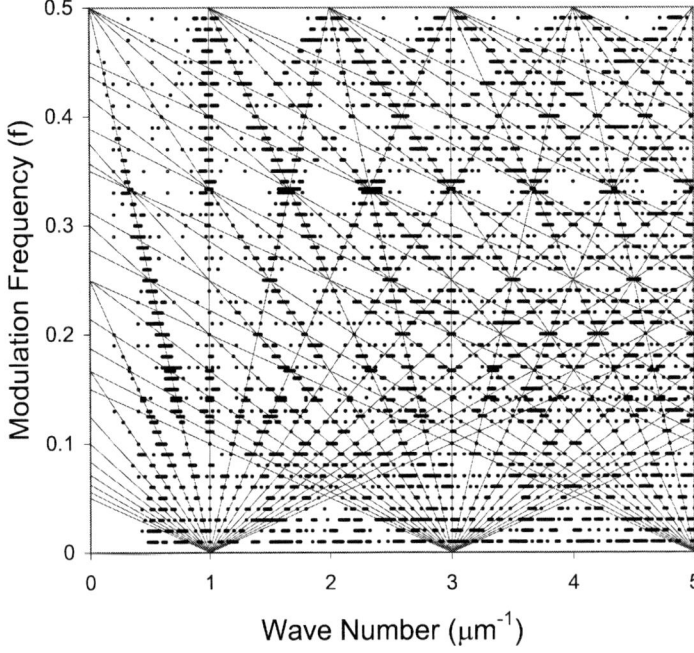

Figure D4 Spectral positions of all stopbands for a modulation amplitude of 1.0. For all designs, the refractive indices for the ambient, substrate, and two films are 1.0, 1.46, 1.45, and 2.0, respectively.

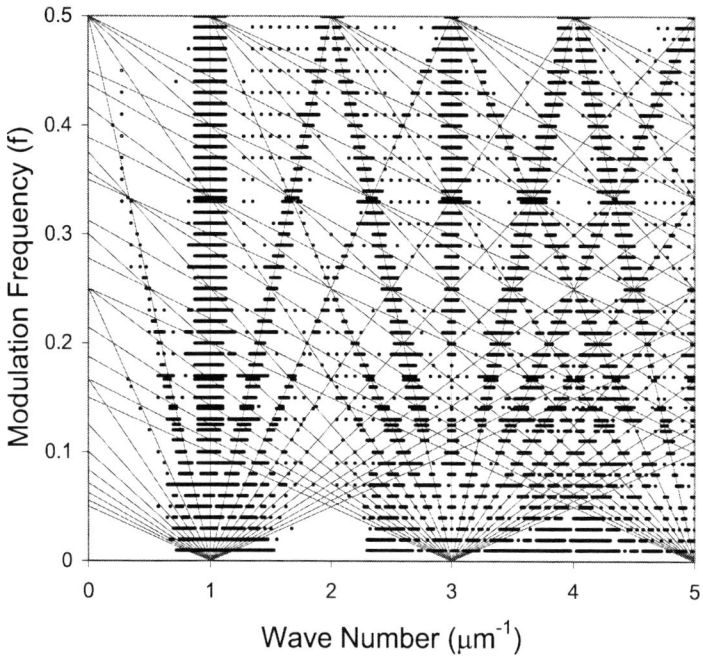

Figure D5 Spectral positions of all stopbands for a modulation amplitude of 0.25. For all designs, the refractive indices for the ambient, substrate, and two films are 1.0, 2.46, 1.4, and 2.2, respectively.

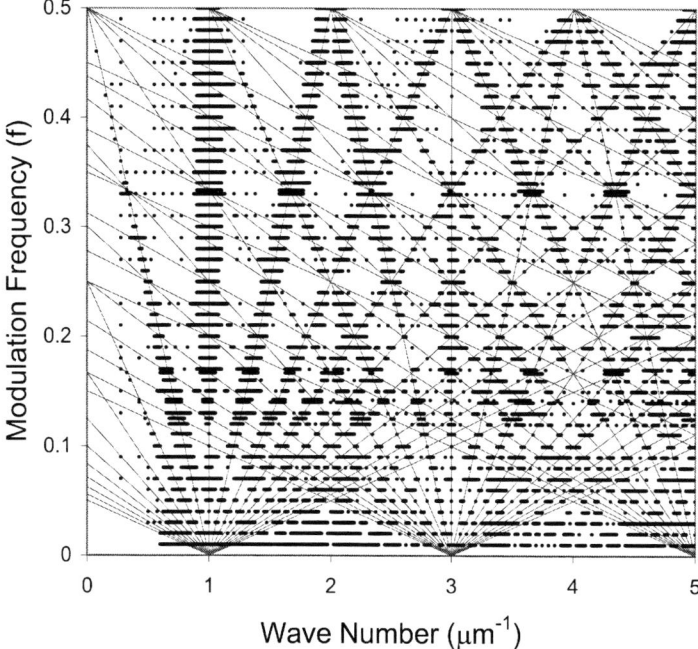

Figure D6 Spectral positions of all stopbands for a modulation amplitude of 0.5. For all designs, the refractive indices for the ambient, substrate, and two films are 1.0, 2.46, 1.4, and 2.2, respectively.

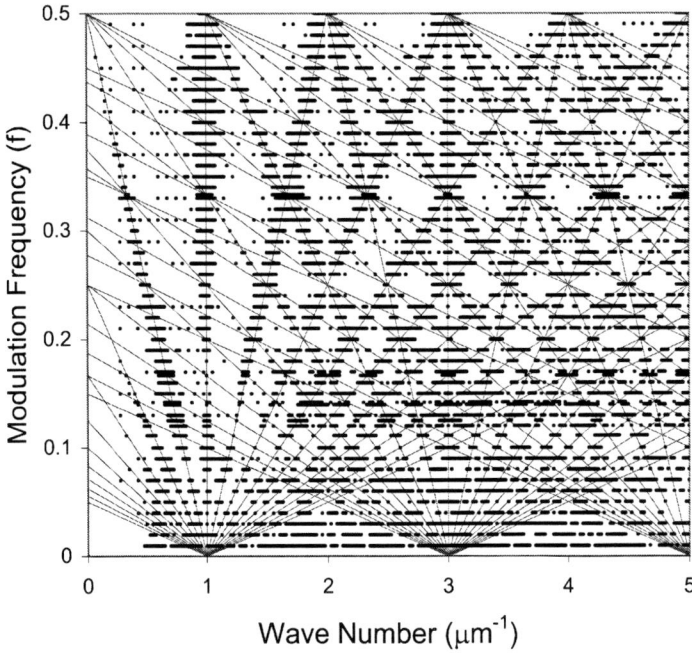

Figure D7 Spectral positions of all stopbands for a modulation amplitude of 0.75. For all designs, the refractive indices for the ambient, substrate, and two films are 1.0, 2.46, 1.4, and 2.2, respectively.

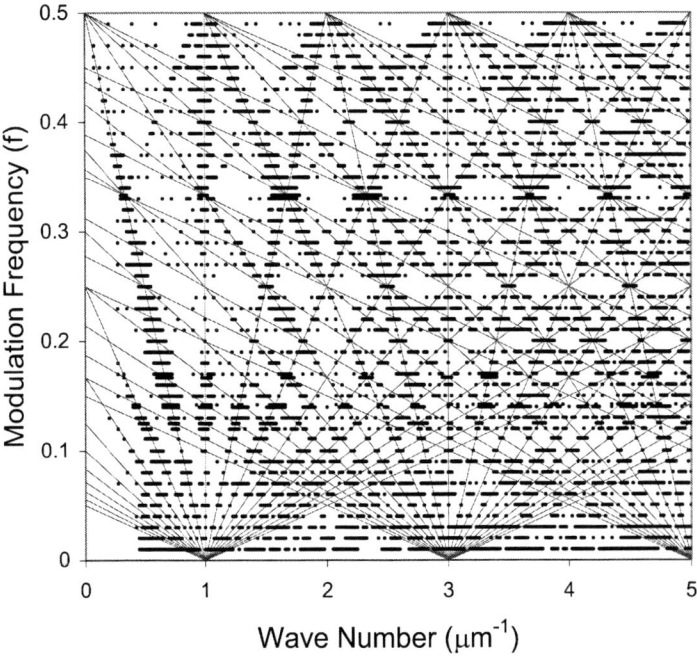

Figure D8 Spectral positions of all stopbands for a modulation amplitude of 1.0. For all designs, the refractive indices for the ambient, substrate, and two films are 1.0, 2.46, 1.4, and 2.2, respectively.

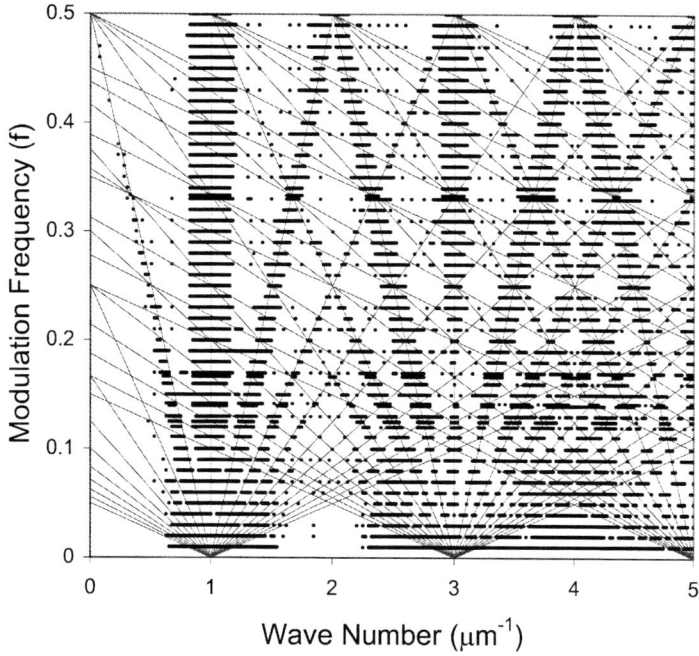

Figure D9 Spectral positions of all stopbands for a modulation amplitude of 0.25. For all designs, the refractive indices for the ambient, substrate, and two films are 1.0, 4.0, 2.2, and 4.0, respectively.

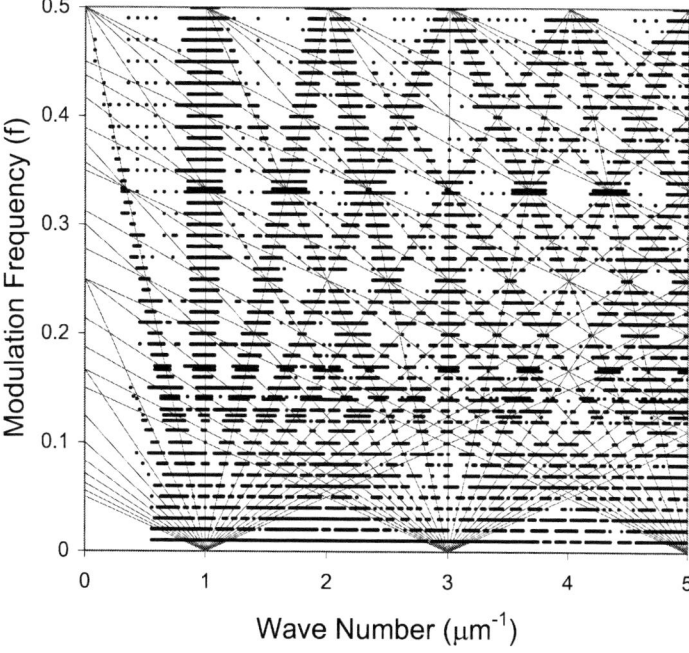

Figure D10 Spectral positions of all stopbands for a modulation amplitude of 0.5. For all designs, the refractive indices for the ambient, substrate, and two films are 1.0, 4.0, 2.2, and 4.0, respectively.

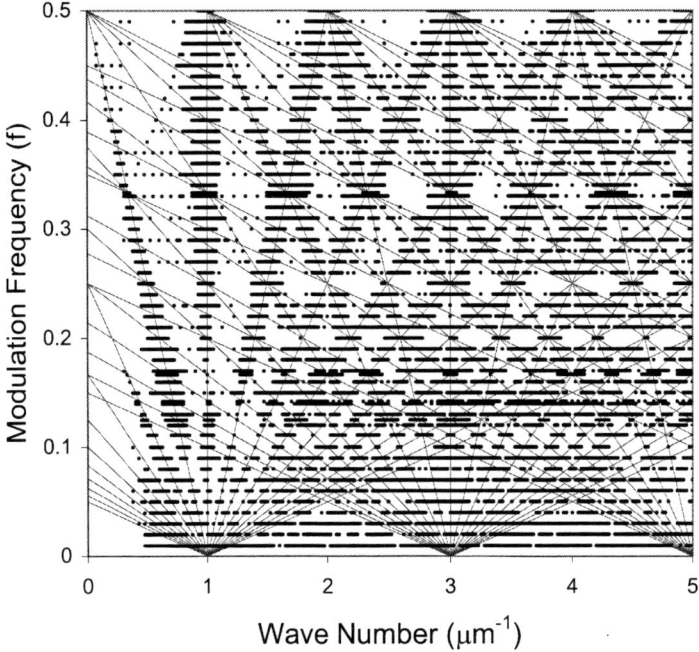

Figure D11 Spectral positions of all stopbands for a modulation amplitude of 0.75. For all designs, the refractive indices for the ambient, substrate, and two films are 1.0, 4.0, 2.2, and 4.0, respectively.

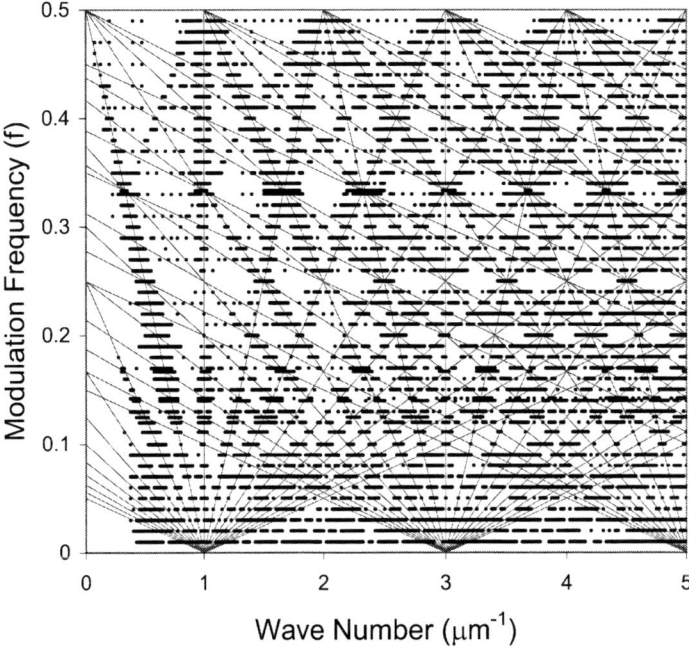

Figure D12 Spectral positions of all stopbands for a modulation amplitude of 1.0. For all designs, the refractive indices for the ambient, substrate, and two films are 1.0, 4.0, 2.2, and 4.0, respectively.

Appendix E

Summary of TMD and LOST Equations

Description	Equations	Eq. No.
TMD	$T(L) = T_{AVG}[1 + k \sin(2\pi fL + \phi')]$	2.14
	$T(L) = T_{AVG}[1 + k \cos(2\pi fL)]$	2.15
(stopband positions)	$R_{AB} = \dfrac{\sigma_B}{\sigma_A} = 2f + 1$	2.21
	$f = \dfrac{1}{2}(R_{AB} - 1)$	2.22
USE-TMD	$\sigma_{M,N} = \sigma_0[2Nf_m + (2M - 1)]$	2.23
(dispersion corrected)	$\sigma_{M,N} \cong \dfrac{\sigma_0[2Nf_m + (2M - 1)]}{1 + \dfrac{1}{2}\left(\dfrac{\Delta n_L}{n_{L_0}} + \dfrac{\Delta n_H}{n_{H_0}}\right)}$	2.24
A-TMD	$T(L) = T_{AVG}[1 + k' \sin(2\pi f_1 L)\cos(2\pi fL)]$	3.3
(apodized)	basic envelope function: $f_1 = \dfrac{1}{2L_{TOT}}$	3.5
	$\sigma_{M,N} \approx \sigma_0[2N(f - f_1) + (2M - 1)]$	3.6
G-TMD (Gaussian)	$m(L) = e^{-B(L - \frac{L_{TOT}}{2})^2}$	3.7
C-TMD	$T(L) = T_{AVG}[1 + k \cos(2\pi fL)]$	4.1
(chirped)	$f(L) = f_b + k_c(L - 1)$	4.2
	$k_c = \dfrac{f_{b'} - f_b}{L_{TOT} - 1}$	4.3
H-TMD (half)	For even-numbered layers: $T(L) = T_{AVG}[1 + k \cos(2\pi fL)]$	4.4
	For odd-numbered layers: $T(L) = T_{AVG}$	4.5
LOST	$T_{U_f}^- = \dfrac{T_{QW1}U_f}{U}(1 - k)$	5.5
	$T_{U_f}^+ = \dfrac{T_{QW1}U_f}{U}\left(1 + \dfrac{S^-}{S^+}k\right)$	5.6

References

1. See, for example, W. H. Hayt, Jr., "Time-varying fields and Maxwell's equations," Chap. 10, and "The uniform plane wave," Chap. 11, in *Engineering Electromagnetics*, 4th ed., McGraw-Hill, New York, New York (1981).
2. M. Born and E. Wolfe, *Principles of Optics*, 5th ed., Pergamon Press, Elmsford, New York (1975).
3. J. D. Rancourt, *Optical Thin Films: User Handbook*, SPIE, Bellingham, Washington (1996).
4. P. Baumeister, *Optical Coating Technology*, Short Course, University of California at Los Angeles, 13–17 January 1997, P. Baumeister, Sebastopol, California (1997).
5. H. A. Macleod, *Thin-Film Optical Filters*, 2nd ed., Macmillan Publishing Company, New York, New York (1986).
6. T. Tamir and H. L. Bertoni, "Lateral displacement of optical beams at multilayered and periodic structures," *J. Opt. Soc. Am.* **61**, pp. 1397–1413 (1971).
7. E. Spiller, "Phase conventions in thin film optics and ellipsometry," *Applied Optics* **23**, p. 3036 (1984).
8. For discussion on classical thin-film designs, see A. Thelen's Introduction, Chap. 1 of *Design of Optical Interference Coatings*, McGraw-Hill, Inc., New York, New York (1989).
9. H. M. Liddell, "Design of filters by analytical techniques," Chap. 2, pp. 46–47, in *Computer-Aided Techniques for the Design of Multilayer Filters*, Hilger, Bristol, UK (1981).
10. See, for example, T. I. Oh, "Infrared minus-filter coatings: design and production," *Applied Optics* **30**, pp. 4565–4573 (1991). ✔
11. P. Baumeister, "Starting designs for the computer optimization of optical coatings," *Applied Optics* **34**, pp. 4835–4843 (1995). ✔
12. R. Jacobsson and J. O. Martensson, "Evaporated inhomogeneous thin films," *Applied Optics* **5**, pp. 29–34 (1966).
13. B. G. Bovard, "Derivation on a matrix describing a rugate dielectric thin film," *Applied Optics* **27**, pp. 1998–2005 (1988).
14. For a discussion of modeling of an inhomogenous layer via subdivided homogeneous ones, see W. H. Southwell, "Use of gradient index spectral filters," *Solid-State Optical Control Devices*, P. Yeh, Ed., *Proc. SPIE* **464**, pp. 110–114 (1984).
15. W. E. Johnson and R. L. Crane, "Introduction to rugate filter technology," *Inhomogeneous and Quasi-Inhomogeneous Optical Coatings*, J. A. Dobrowolski and P. G. Verly, Eds., *Proc. SPIE* **2046**, pp. 88–108 (1993).
16. W. H. Southwell, "Using apodization functions to reduce sidelobes in rugate filters," *Applied Optics* **28**, pp. 5091–5094 (1989).
17. W. H. Southwell, "Using wavelets to design gradient-index interference coatings," *Inhomogeneous and Quasi-Inhomogeneous Optical Coatings*, J. A. Dobrowolski and P. G. Verly, Eds., *Proc. SPIE* **2046**, pp. 46–59 (1993).

18. P. G. Verly, "Design of inhomogeneous and quasi-inhomogeneous optical coatings at the NRC," *Inhomogeneous and Quasi-Inhomogeneous Optical Coatings*, J. A. Dobrowolski and P. G. Verly, Eds., *Proc. SPIE* **2046**, pp. 36–45 (1993).

19. B. G. Bovard, "Rugate filter design: the modified Fourier transform technique," *Applied Optics* **29**, pp. 24–30 (1990).

20. P. G. Verly, J. A. Dobrowolski, W. J. Wild, and R. L. Burton, "Synthesis of high rejection filters with the Fourier transform method," *Applied Optics* **28**, pp. 2864–2875 (1989).

21. Private correspondence with P. G. Verly.

22. A. Thelen, *Design of Optical Interference Coatings*, McGraw-Hill, New York, New York (1989).

23. L. Epstein, "Improvements in heat-reflecting filters," *J. Opt. Soc. Am.* **45**, pp. 360–362 (1955).

24. B. E. Perilloux, "Discrete thin-film layer thickness modulation," *Applied Optics* **37**, pp. 3527–3532 (1998).

25. M. Thomsen and Z. L. Wu, "Polarizing and reflective coatings based on half-wave layer pairs," *Applied Optics* **36**, pp. 307–313 (1997).

26. H. M. Liddell, *Computer-aided Techniques for the Design of Multilayer Filters*, Hilger, Bristol, UK (1981).

27. B. E. Perilloux, "Discrete thin-film thickness-modulated designs: spacing of all possible stopbands," *Applied Optics* **38**, pp. 2911–2915 (1999).

28. J. H. Apfel, "Optical coating design with reduced electric field intensity," *Applied Optics* **16**, pp. 1880–1885 (1977).

29. M. Flannery, E. Loh, Jr., and M. Sparks, "Nearly perfect multilayer dielectric reflectors: theory," *Applied Optics* **18**, pp. 1428–1435 (1979).

30. O. Arnon and P. Baumeister, "Electric field distribution and the reduction of laser damage in multilayers," *Applied Optics* **19**, pp. 1853–1855 (1980).

31. F. Abelès, "Remarque sur l'influence de la dispersion dans les systèmes de couches minces diélectriques," *Journal de Physique et le Radium* **19**, p. 327 (1958).

32. B. E. Perilloux, "Infrared thin film polarization preserving reflectors," pp. 50–70, in *Infrared Thin Films*, Vol. **CR39**, SPIE, Bellingham, Washington (1991).

33. P. G. Verly, "Fourier Transform technique with frequency filtering for optical thin film design," *Applied Optics* **34**, pp. 688–694 (1995).

34. For the definition of GDD, see, for example, R. Fork, C. Cruz, P. Becker, and C. Shank, "Compression of optical pulses to six femtoseconds by using cubic phase compensation," *Optics Letters* **12**, pp. 483–485 (1987).

35. R. Szipöcs, K. Ferencz, C. Spielmann, and F. Krausz, "Chirped multilayer coatings for broadband dispersion control in femtosecond lasers," *Optics Letters* **19**, pp. 201–203 (1994).

36. G. Tempea, F. Krausz, C. Spielmann, and K. Ferencz, "Dispersion control over 150 THz with chirped dielectric mirrors," *IEEE J. of Selected Topics in Quantum Electronics* **4**, pp. 193–196 (1998).

37. N. Matuschek, F. X. Kärtner, and U. Keller, "Theory of double-chirped mirrors," *IEEE J. of Selected Topics in Quantum Electronics* **4**, pp. 197–208 (1998).

38. W. H. Southwell, "Extended-bandwidth reflector designs by using wavelets," *Applied Optics* **36**, pp. 314–318 (1997).

Index

Bruce E. Perilloux is an engineering manager at the Photonics Group, Optics Business Unit, of Coherent, Inc. in Auburn, California. For most of his career at Coherent, he has held the position of thin-film R&D engineer. He started at Coherent in 1985 after receiving a BS in electrical engineering and an MS in engineering (thin films, ellipsometry, electro-optics), both at the University of New Orleans. He is currently an editing referee for the journals *Applied Optics*, *Journal of the Optical Society of America*, and *Optical Engineering*; the author of 14 technical papers on thin films and electro-optics; and the holder of six  optical thin-film design patents. Recently, he has worked on the design and the manufacturing process for multiple-wavelength coatings, broadband coatings for the camera optics of the KECK II telescope, and the development of ion plasma coating methods. He is a member of OSA, SPIE, Sigma Xi, and Phi Kappa Phi.